CHEMISTRY
AND CRIME

ABOUT THE EDITOR

Sᴀᴍᴜᴇʟ M. GERBER IS A SPECIALIST in the chemistry and technology of dyes and their intermediates. He received a B.S. degree in
chemistry from the City College of New York and an M.S. and Ph.D.
from Columbia University. Most of his professional career was with the
American Cyanamid Company, where his assignments included Chief
Chemist of Dyes and Intermediate Manufacturing and Manager of Dyes
and Chemicals R&D. At present, he is a consultant in the field of dyes
and related products. His professional publications and patents are largely
in the field of dyes and intermediates. They include an extensive com-
pilation of Soviet contributions on azo and diazo chemistry.

Dr. Gerber's interest in forensic chemistry originated with Sherlock Holmes.
He is a member of the Baker Street Irregulars and related groups. His
lecture, "Sherlock Holmes, Chemist," has been widely presented. He is
married to Nancy Nichols Gerber, a chemist at Rutgers' Waksman Institute, and they have two children and one grandchild.

ABOUT THE AUTHORS

ELY M. LIEBOW IS ASSOCIATE PROFESSOR OF ENGLISH at Northeastern Illinois University and former chairman of the department. He did his undergraduate work in English at American University and pursued graduate studies at the University of Chicago and Rutgers University. Long a devotee of Sherlock Holmes and detective fiction in general, he has written articles and delivered papers on such diverse worthies as Rabbi Small, Father Brown, Joyce Porter's Dover, and Sherlock Holmes himself. He is a former Sir Hugo (President) of Hugo's Companions, the Chicago Scion Society of the Baker Street Irregulars, and is an investitured member of the BSI. He also serves as editorial advisor to the *Baker Street Miscellanea*. He was invited to address the world's only Henry Fielding Society (the novelist-magistrate being the true founder of the British Police Force), located in Her Majesty's Royal Police College, Hampshire, England, and is a member of the society. He was also asked to speak at New Scotland Yard. His biography, "Dr. Joe Bell: Model for Sherlock Holmes," was published in 1982. He is coauthor of "Write It Right," a college grammar; and coeditor of "Age: A Work of Art," a literary anthology dealing with the aged.

NATALIE FOSTER IS AN INSTRUCTOR in the chemistry department of Lehigh University. She holds both the Doctor of Arts and the Ph.D. in chemistry from Lehigh and is currently doing research in the field of radiopharmaceuticals for detection of malignancies. Her publications include methods of radiohalogenating organic molecules, the use of porphyrins as imaging agents, and several in the field of the history of medicine on homeopathy and early pharmaceutical companies. She is a member of the American Chemical Society, the American Society for Photobiology, Sigma Xi, and the Red-Headed League, a scion society of the Baker Street Irregulars.

RICHARD SAFERSTEIN IS CHIEF FORENSIC CHEMIST with the New Jersey State Police Laboratory. He received his B.S. and M.A. degrees in chemistry from City College of the City University of New York and his Ph.D. in chemistry from City University of New York. He was a forensic chemist with the U.S. Treasury Department and an analytical chemist with Shell Chemical Company. He currently serves on the editorial board of the *Journal of Analytical and Applied Pyrolysis*. He has written numerous technical publications and "Criminalistics: An Introduction to Forensic Science" and has edited the "Forensic Science Handbook."

PETER R. DE FOREST IS A PROFESSOR OF CRIMINALISTICS at the John Jay College of Criminal Justice of the City University of New York. He became interested in the field of forensic science in 1960 when he obtained a part-time job in a crime laboratory while he was still a chemistry undergraduate. After two years of working part time and taking a chemistry curriculum, he transferred to the University of California at Berkeley to complete his B.S. degree in criminalistics. After earning his doctorate from Berkeley in 1969, he came to New York as an assistant professor to teach in the B.S. and M.S. programs in forensic science at the John Jay College. He has published a number of articles in several journals and recently coauthored a textbook entitled "Forensic Science— An Introduction to Criminalistics." He recently contributed an extensive chapter on forensic microscopy to "Forensic Science Handbook," edited by Richard Saferstein.

VINCENT P. GUINN IS PROFESSOR OF CHEMISTRY at the University of California at Irvine. He took A.B. and M.S. degrees in chemistry at the University of Southern California, and then a Ph.D. in physical chemistry at Harvard University. Shortly thereafter, he took training in radiochemistry at the Oak Ridge Institute of Nuclear Studies. From 1949 to 1961, he was a research chemist with the Shell Development Company, becoming supervisor of Radiochemistry in 1956. From 1962 to 1970, he was technical director of the Activation Analysis Program at the General Atomic Company. At the University of California, he now does research and teaches in the areas of neutron activation analysis and forensic chemistry. His publications in these fields total over 200. Professor Guinn is a Fellow of the American Academy of Forensic Sciences and a Fellow of the American Nuclear Society. He is also a member of the American Chemical Society, the Forensic Science Society (of England), and the California Association of Criminalists. He frequently testifies in court as an expert witness. In 1964 he received the Special Award of the American Nuclear Society, and in 1979 he received the George Hevesy Medal (in radioanalytical chemistry).

FRANCES M. GDOWSKI IS PRINCIPAL FORENSIC CHEMIST with the New Jersey State Police Laboratory. She received her B.S. in Biology from Wilmington College (Ohio) and is currently enrolled in the M.S. program at the John Jay College of Criminal Justice of the City University of New York. She has taken training in bloodstain analysis from the Federal Bureau of Investigation, the University of Pittsburgh, and John Jay College. She was one of five persons selected to evaluate a bloodstain analysis program sponsored by the Law Enforcement Assistants Association. She has presented numerous papers at forensic science meetings, and conducted seminars for forensic chemists. Currently, she gives training in bloodstain and semen analysis and presents lectures to police investigators and other legal specialists.

LAWRENCE KOBILINSKY IS ASSISTANT PROFESSOR of biology and immunology at John Jay College of Criminal Justice of the City University of New York and a member of the doctoral faculty in the Ph.D. program in biochemistry at the City University of New York. He received his B.S., M.A., and Ph.D. degrees in the field of biology from the City College of the City University of New York. He was a research fellow, research associate, and visiting investigator at the Sloan–Kettering Institute for Cancer Research. His areas of specialization are immunochemistry and forensic serology. He serves as a forensic science consultant and is court qualified. He has reviewed books and symposia and has published in the fields of forensic science and in the immunochemistry of retroviral infection.

JOSEPH L. PETERSON IS THE DIRECTOR of the Center for Research in Law and Justice at the University of Illinois at Chicago, where he is also Associate Professor of Criminal Justice. He received his doctorate in criminology from the University of California at Berkeley where he specialized in forensic science. He was Forensic Science Program Manager of the National Institute of Law Enforcement and Criminal Justice for three years. Dr. Peterson formerly directed the Criminal Justice Center at John Jay College of Criminal Justice of the City University of New York and the Forensic Sciences Foundation, Inc., in Rockville, Maryland. He has authored several articles and research monographs in the field of forensic science and is the editor of "Forensic Science: Scientific Investigation in Criminal Justice."

NICHOLAS PETRACO HAS WORKED AS A DETECTIVE in the Crime Laboratory of the New York Police Department for the past 15 years. Most of his work involves microscopy and trace examinations of hair and fiber. He also evaluates particulate matter and physiological fluids. He uses evidence to reconstruct crimes to prove guilt or innocence. He is an instructor at the John Jay College of Criminal Justice of the City University of New York in forensic science and general criminalistics. He is a Fellow of the American Academy of Forensic Science and a member of the New York Forensic Microscopy Society. He has served in court frequently as an expert witness and has authored several publications.

CONTENTS

PREFACE

Holmesians are committed to maintaining the myth that Sherlock Holmes was not a myth. Arthur Conan Doyle's writings are full of ideas on forensic chemistry; articles have been written on Holmes' chemical contributions; moreover, Watson himself said that Holmes' knowledge of chemistry was profound.

Natalie Foster of Lehigh University, who has a keen interest in detective fiction, and I thought it would be a great idea to have a symposium on Chemistry in Crime—Fact and Fiction. We incorporated the fictional and forensic parts and noted, in many cases, a relationship between the two. The first three chapters clearly demonstrate this connection. The remaining chapters present descriptions of modern techniques in various subdisciplines of forensic science.

I want to thank all the contributors and the members of the American Chemical Society Books Department, especially Suzanne B. Roethel, Janet S. Dodd, Paula M. Bérard, and Anne G. Bigler.

A special debt of gratitude is due to Richard Saferstein of the New Jersey State Police.

Samuel M. Gerber
May 1983

FROM SHERLOCK HOLMES

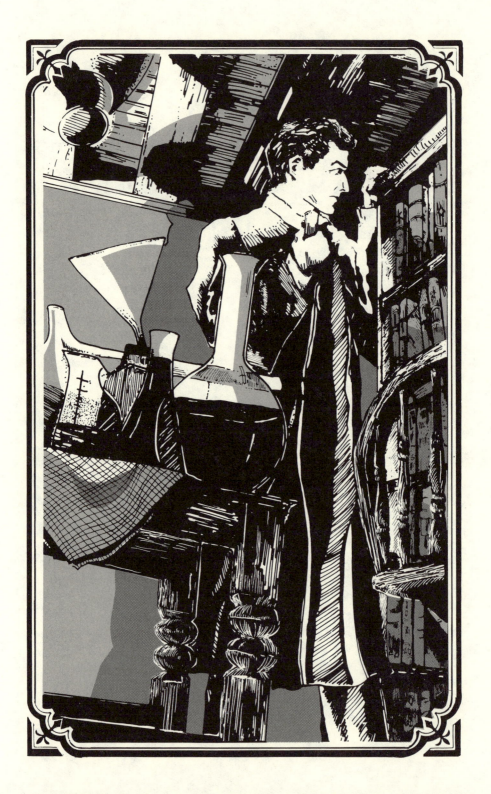

1

Medical School Influences on the Fiction of Arthur Conan Doyle

Ely Liebow

Arthur Conan Doyle derived many benefits from his Edinburgh upbringing, his medical training, and his professors of medicine and chemistry. Doyle's own autobiography and fiction attest to strong influences, as do various other sources.

A school or a city can, of course, have diverse and subtle effects upon a person, especially an inveterate romantic. Wordsworth sang his nature songs to the Lake District, finding sermons under every rock. Wellington felt that the Battle of Waterloo was won on the playing fields of Eton, but E. M. Forster, deeply seared by the memories of his days in England's "public" schools, felt they fashioned the overdeveloped body and the underdeveloped heart.

Upbringing in Edinburgh

Arthur Conan Doyle's Edinburgh may not have been exactly what Dublin was to James Joyce, but in a curious sense there are many similarities. As disaffected young Catholics, each left the city of his boyhood, never to return. Joyce, of course, wrote about nothing but Dublin the rest of his life, but Doyle the writer is associated with London. Like Joyce, Arthur Conan Doyle never forgot his native city, and had a romantic love–hate relationship with it. It appears over and over again in his novels and short stories, and twice he went back to run for Parliament from Edinburgh.

Unlike the Dubliner, Doyle grew up in a sternly Protestant city and had relatively little to say about his youthful friends and companions. His Irish family was just as devout as the Joyce clan, and when he fell away from the Church they were just as grieved. Shortly after disavowing Ca-

0784/83/0003$06.00/0
© American Chemical Society

tholicism, and barely eighteen years old, he enrolled in Edinburgh University's famed medical school. In one way or another he thought about his medical training and wrote about it for the rest of his life.

It was young Doyle's mother who suggested not only that he become a doctor, but that he enroll in what was then (1876) the world's finest medical school, the University of Edinburgh. Thus, unwittingly, it was the ma'am (as Doyle always referred to his mother) who supplied him with both of his professions, for his mother read to him, instructed him in heraldry, and filled his mind with romance, gallantry, and literature. Medicine and the formidable pile of stone that represented the university may have been the furthest thing from the young man's mind. Certainly, all his fondest and most vivid recollections of his earlier schooling centered around sports, newspapers, and his favorite authors.

Young Doyle applied for a bursary (scholarship), the Grierson Bursary. It was worth £40, an enormous sum for a young man with no income and from a desperately poor home. The bursary came through; everyone at the Doyle household rejoiced; when young Arthur paid his "greasy £1 note" to the bursar to enroll, he learned that the bursary had been withdrawn at the last minute. He was a fighter, however, and he needed the money. Somehow the school scraped up a £7 scholarship.

Years in Medical School

In some ways Edinburgh, especially Doyle's medical school training, became part and parcel of his being. The school helped shape his scientific and literary curiosity; introduced him to different aspects of sports; and instilled a large measure of discipline in him and many of his classmates; and it was certainly here that he developed an interest in chemistry, drugs, and the laboratory.

Arthur Conan Doyle was not very romantic about his days as a hopeful medical student. He would be far more nostalgic about the life of a young graduate in the medical fraternity. The professors at the university seldom deigned to see, counsel, or mingle with the students out of class. Doyle stressed the hauteur and unavailability of the medical school professors in his neglected early novel *The Firm of Girdlestone*. By the time Doyle became a medical student, university students could take approximately half their classes at other recognized schools and still receive credit. Joe Bell, as he was known affectionately throughout Edinburgh, and a host of celebrated medical men were classified as extramural teachers in the Royal Infirmary of Edinburgh when Doyle enrolled. Doyle, along with most of his classmates, took a good number of extramural classes.

Influence of Joe Bell

Joe Bell chose young Doyle to be his dresser (assistant) at the end of Doyle's second year. Doyle wondered what Joe Bell saw in him. Perhaps this was modesty, but in truth Joe Bell watched his students with, as one student described it, "the look of eagles." "Doyle was always making notes," Bell told a *Pall Mall Gazette* reporter. "He seemed to want to copy down every word I said. Many times after the patient departed my office, he would ask me to repeat my observations so that he would be certain he had them correctly." Later, he told the same reporter, "I always regarded him as one of the best students I ever had. He was exceedingly interested always upon anything connected with diagnosis, and was never tired of trying to discover those little details which one looks for."

All Sherlockians know that on that fateful night in Southsea in 1886 when the struggling young doctor sat down to write his first detective story, he recalled the fact, the voice, the method of "my old mentor, Joe Bell." It was the method that was to become the touchstone, the magic of Sherlock Holmes.

Bell wrote:

> I recollect one time when a patient walked in and sat down. "Good morning, Pat," I said, for it was impossible not to see that he was an Irishman.
>
> "Good morning, your honor," replied my patient.
>
> "Did you like your walk over the links today as you came in from the south side of town?" I asked.
>
> "Yes," said Pat. "But the divil. Did your honor see me?"
>
> Well, Conan Doyle could not see that, absurdly simple as it was. On a showery day, as that had been, the reddish clay at bare parts of the links adheres to the boot, but a tiny part is bound to remain. There is no such clay anywhere else around the town for miles.
>
> Once the patient was gone. . . Conan Doyle made me explain about the boots and clay, and he wrote my every word down in his little book.

In Doyle's own "The Five Orange Pips" Holmes tells his young client whom he has just met for the first time, "You have come up from the southwest, I see."

"Yes, from Horsham."

"That clay and chalk mixture which I see upon your toecaps is quite distinctive."

How did Bell start the game, the Method? Speaking in his "Rich Scots" he would pass around the class a vial filled with amber-colored liquid. In a voice of subdued humor ("we never knew how much he kept that sharp tongue in his cheek"), he would point out to each new class that "This, gentlemen, contains a most potent drug. It is extremely bitter to the taste. Now I wish to see how many of you have developed the powers of observation that God granted you. 'But sair,' ye will say, 'it can be analyzed chemically.' Aye, aye, but I want you to taste it—by smell and taste. What! You shrink back? As I don't ask anything of my students which I wouldn't do alone wi' myself, I will taste it before passing it around." He was obviously enjoying himself.

He would then dip a finger in the liquid, put his finger in his mouth, suck it, and then grimace. "Now you do likewise," and he would pass the liquid around. Each student would taste the harsh concoction, make a face, and pass the awful stuff to his neighbor. When the vial finally came back to him, Bell would look out over the pinched faces, and slowly start to chuckle. "Gentlemen, gentlemen," he would coo, "I am deeply grieved to find that not one of you has developed his power of perception, the faculty of observation which I speak so much of, for if you had truly obsairved me, you would have seen that, while I placed my index finger in the awful brew, it was the middle finger—aye—which somehow found its way into my mouth."

Bell also recounted one of Doyle's favorite cases to the *Pall Mall Gazette* reporter. He told of the patient who walked into the room where he was instructing students. "He has been a soldier in a Highland regiment, and probably a bandsman," Bell stated. Bell pointed out the swagger in his walk, suggestive of the piper, and deduced from his shortness that he had to be a bandsman. When asked about his station in life, the man said he was a cobbler and had never been in the army in his life.

"This was rather a floorer, but being absolutely certain I was right, and seeing something was up, I did a pretty cool thing. I told two of the strongest clerks to remove the man to a side room. . . . I went and had him stripped, and I daresay your own acuteness has told you the sequel."

"I'm dashed if I do," said the Watson-like reporter.

"Why," said Dr. Bell, "under the left breast I instantly detected a little blue 'D' branded on his skin. He was a deserter. That was how they used to mark them in the Crimean days."

Doyle was also well aware that Bell had outfitted a small laboratory on his ward at the Royal Infirmary and built a superb laboratory in his last two residences in Edinburgh. Bell was an experimenter. In his article "Looking Back: 1907–1860," Dr. John Chiene recalls that whenever he came onto Ward XI there would be Joe Bell, intent over a bunsen lamp [sic]. Indeed, when Watson first saw Holmes he was bending over a table "which bristled with retorts, test-tubes, and little bunsen lamps."

Joe Bell also told his dressers to learn the Scottish idiom, the vernacular. Doyle, who did indeed absorb all that his canny mentor taught him, did not think this advice was of utmost importance. Then, as he recounts, sure enough an Edinburgh farrier came up to him in the out-patient clinic complaining of a "bealing in the oxter."

"This beat me," said Doyle, and he was told that the patient had an abscess in the armpit. Thus, when Dr. Bell said, on first observing and listening to a patient, that he was a bellringer from south of the Tweed, it had as much to do with dialect and idiom as with observation and general knowledge. "I had ample chance of studying his methods and of noticing that he often learned more of his patient by a few quick glances than I had done by my questions," said Doyle.

Just how did Joe Bell influence, aid, or abet Doyle the writer? Al-though Doyle said that Bell gave him some plots for grisly stories, he also declared that he never used any of the plots suggested. Adrian Conan Doyle also claimed that Doyle didn't use any of Bell's ideas. He is probably right, because the quiet, classical Joe Bell had some ideas for which the world was not yet ready. Bell probably anticipated modern surgical scrub-up procedures, penicillin, and the autoclaving of instruments, and he may have also anticipated (indirectly, it must be admitted) the Tylenol murders of 1982—on a literary level. "Poison a well at Mecca," he told the *Strand* readers, "with the cholera bacillus, and the holy water which the pilgrims carry off in their bottles will infect a continent, and the rags of the victims of the plague will terrify every seaport in Christendom." As Irving Wallace put it: "Dr. Bell suggested in 1892 that Holmes pit himself against a germ murderer, and hinted at knowledge of one such case. Doyle was quick to question if a bacteriological killer might not be too complex for the average reader." In all likelihood, Doyle, who could work up a splendid chiller of a ghost story, didn't want his mysteries to be too real.

Influence of Crum Brown

While he may have used Joe Bell as a model for his most famous literary creation, Doyle had nearly as much to say about Crum Brown, professor of chemistry, as any of the other instructors.

Crum Brown was a scholarly absent-minded soul, one of the most popular teachers at the university. In his autobiography, Doyle recalled the kindly manner of the man, and how the students looked forward to his lab experiments (and the number of experiments that went awry). He was one of the few instructors willing to teach chemistry to a small band of women medical students (1869–1874), and they remembered how he taught the usual chemistry courses, but preferred the prize competitions, how he stressed original and unusual research.

In his diary, Doyle the young doctor (soon to be a writer), vowed he would write significant articles for the medical journals (he wrote a few) and also averred that he would put into effect the spirit of experi-mentation that he learned in classes at the university. While doing research for his dissertation (a paper dealing essentially with tabes dorsalis, a strain of syphilis), he tells us he was dealing with "remedial drugs. . . . I myself have taken as many as forty minims of Murrell's Solution without inconvenience." Brown's colleagues, A. C. Curor and John Chiene, stated that Crum Brown often told the young students of the drugs with which he experimented.

Doyle's first published medical article appeared in the prestigious *British Medical Journal* in 1879, when he was a third-year medical student. The article, "Gelsiminum as a Poison," is yet another account of his experiments with drugs on his own body. Certainly the same fascination with drugs and self-experimentation are to be found in *A Study in Scarlet*, "Adventure of the Devil's Foot," *The Parasite*, and other works.

It also seems fairly obvious from Doyle's story, "The Doctors of Hoyland," that he learned well Crum Brown's lessons of true chemical experiments. The story "The Doctors of Hoyland" is one of Doyle's most autobiographical tales and certainly one of the most interesting. Supposedly Conan Doyle was vigorously opposed to the Suffragette movement. None of his medical school classmates (1876–1881) was female, but "The Doctors of Hoyland" appeared in 1894. In it, one Dr. James Ripley, fresh from the University of Edinburgh Medical School, comes to practice in a small English town. Because of his newer methods and scientific spirit he easily supplants the two old practitioners in the hearts of the natives. For five or six years he reigns unchallenged, and then—to his horror—hears of a "celebrated Dr. Verrinder Smith," newly arrived in town. Dr. Smith was educated and graduated with distinction at Edinburgh; then went on to Paris, Berlin, and Vienna, culminating with the doctor's winning the famed Lee Hopkins Award. Overcoming his first reaction of fear and jealousy, the good doctor calls on his new colleague and is let in by "a little woman with shrewd, humorous eyes, holding a pince-nez in her hand." Asking for Dr. Smith, he learns he is talking to Dr. Smith. Upon gleaning this information, Dr. Ripley is genuinely shocked. "He had never seen a woman doctor before, and his whole conservative soul rose up in revolt at the idea. He could not recall any Biblical injunction that man should remain ever the doctor and the woman the nurse, and yet he felt as if a blasphemy had been committed. . . .a man, of course, could come through such an ordeal with all his purity, but it was nothing short of shameless in a woman." Barely able to tell her that her true place is in the kitchen (or the bedroom), he is downright rude to her, and he knows it: "She had been courteous; he had been rude."

In a fairly adroit manner, Doyle has the young lady save the doctor's leg some months later after an accident on the road. We see him not only change his mind about her, but also preferring her ministrations to those of his brother, a celebrated London surgeon. Finally, of course, he proposes to her—she whom he had thought of as "an unsexed woman."

" 'What, and unite the practises?' said she, smiling." She apologizes for ribbing him (he is very stuffy), and finally tells him she intends to take an opening in the Paris Physiological Laboratory. And so there was again only one doctor in Hoyland, and as Doyle wraps things up, folks could see that "Dr. Ripley had aged many years in a few months. . .and that he was less concerned than ever with the young ladies whom chance, or their careful country mommas, placed his way."

Now, consciously, and perhaps unconsciously, Doyle is bringing to bear more about the medical school than the uninformed reader could possibly ascertain. Dr. Smith won a hotly contested chemistry award at Edinburgh; she is extremely spunky and independent and not very feminine in her bearing. The young author (the stories were all written before he was 33) is weaving several strands of a large, loose tapestry.

Undoubtedly Doyle had in mind a dynamic young lady whose life was only recently (1980) featured in an eight-part BBC production: Sophia Jex-Blake. After being turned down by Harvard Medical School in 1865, she came back to Britain and tried to "get into the best medical school anywhere—Edinburgh." She was allowed to enter in the summer of 1869, but, being female, naturally she had to take each class alone as there was no mixing of the sexes. Being told that such an arrangement was too costly for the university, the determined Sophia came back with six bright ladies, all of whom were reluctantly admitted. One of them, Grace Pechey, took the equivalent of Dr. Verrinder Smith's Lee Hopkins Award.

Doyle would not have forgotten Ms. Pechey's achievement, for while she "won" the famous Hope Award (named for a chemistry professor before Doyle was enrolled there), it was given to a male student, and the entire city joined the battle. The professor caught in the middle was Crum Brown. Everything else about Verrinder Smith sounds like the rather "man-ish" Sophia. Her brilliant defense of women and her knowledge of women in the history of medicine came straight from Sophia's brilliant courtroom testimony when she took on the university senate from 1870 to 1874. Sophia also left Britain for the continent. In fact, the whole ugly scene probably etched itself on the young man's mind, for we get a prototype of Sophia in at least three other Doyle works, culminating in the dynamic Mrs. Westmacott in *Beyond the City*.

Autobiographical Fiction

The beleaguered women or the women medical-hopefuls theme was but one direct episode or influence of the medical school that found its way into Doyle's fiction. Another *Round the Red Lamp* story, "His First Operation," comes straight from Doyle's early days at the Royal Infirmary of Edinburgh. In this superb sketch, a third-year dresser takes a first-year student to his (the neophyte's) first operation. Doyle, of course, was Dr.

Bell's dresser in the Royal Infirmary's Outpatient Clinic. In his *Leaves from the Life of a Country Doctor,* Clement Gunn, a classmate of Doyle, tells us that the taking of a first-year student to his first operation took on all the ritual of college hazing.

In Doyle's account, on the way to the operating room—with the neophyte's heart in his mouth without benefit of anesthesia—the dresser greets a fellow outpatient clerk.

"Anything good on?"

"You should have been here yesterday. We had a regular field day. A papliteal aneurism, a Colles' fracture, a spina bifida, a tropical abscess, and an elephantiasis. How's that for a single haul?" In his autobiography, Doyle noted that Joe Bell would often handle 70 or 80 cases in an afternoon.

Suddenly they are in the operating amphitheatre: "The tiers of horseshoe benches rising from the floor to the ceiling were already packed, and the novice as he entered saw vague curving lines of faces in front of him."

"This is grand," the senior man whispers. "You'll have a rare view of it all."

When the novice asks about the two men at the operating table, he is told they are dressers. "One has charge of the instruments, and the other of the Puffing Billy. It's Lister's antiseptic spray, you know, and Archer [one of the operating surgeons] is one of the carbolic-acid men. Hayes [another surgeon] is the leader of the cleanliness-and-cold-water school, and they all hate each other like poison."

What the novice is learning is that not all the doctors at the famed medical school followed the great Lister with his new antiseptic technique. Doyle was well aware that Joe Bell was one of the first surgeons who did believe in and use the technique, though he was also one of the first proponents to point out some obvious weaknesses. Doyle was also keenly aware that Dr. James "Dismal Jeemy" Spence led those opposed to the germ theory. In several of the medical stories he points out the intense rivalries among some of the greatest physicians on the staff. While many of the antics of the rivals were petty, he was impressed that the major disagreement stemmed from medical or philosophical principles.

In another *Round the Red Lamp* story, "A False Start," a poor, newly graduated doctor, Horace Wilkinson, is trying to make ends meet. Like Doyle, he often answered the bell himself, and when trying to make a none-too-easy diagnosis, he knew that "some of his old Edinburgh professors would have diagnosed his case by now, and would have electrified the patient by describing his own symptoms before he had said a word about them." Again, shades of Joe Bell.

In some ways the most obvious or direct influence of the medical school upon the young writer appears in "A Medical Document," another of the *Round the Red Lamp* stories. At the end of this story, a young medic has carefully recorded some truly bizarre case histories from a little group of doctors that he (the budding writer) calls the Midland Branch of the British Medical Association. Their bizarre stories are interesting, and one dealing with syphilis also appeared in Doyle's dissertation.

"Behind the Times" is a poignant tale about a figure depicted by Doyle in many of his novels and stories—the old but extremely able family doctor. In Doyle's own life it was a Dr. Hoare of Birmingham for whom he worked during the summer of 1879 while still a student. In several stories Doyle calls him "Dr. Horton," but here he is called, fittingly enough, Dr. Winter. The speaker here is a clever, competent newly graduated doctor who was brought into the world by Dr. Winter. No one knows the old man's age. He is opposed to the use of chloroform. Joe Bell's teacher and colleague Dr. J.B.Y. Simpson perfected chloroform as an anesthesia just about the time that Joe Bell entered medical school. Many of the lightning-fast operators were leery of chloroform, and it was administered by dressers, nurses, and orderlies until World War I. Winter referred to a stethoscope as "a new-fangled French toy," and, as mentioned earlier, "The germ theory of disease set him chuckling for a long time, and his favorite joke in the sick room was to say 'shut the door, or the germs will be getting in.' " Joe Bell wrote that Dismal Jeemy Spence used to shout, always in Lister's presence, "Shut the door; ye'll let the germs oot." Dr. Winter "had more practical knowledge of [dietetics] than any one whom I met." Patrick Heron Watson, "one of Spence's boys," was the recognized expert in dietetics (probably the only one) when Doyle was a medical student.

If we can lend any credence at all to Henri Mutrux's notion that Doyle projected himself into many of his literary creations, then he didn't have a lot of projecting to do while writing "The Croxley Master," a non-"Red Lamp" but very medical story. While it may not be autobiographically accurate, it has many situations taken from Doyle's own medical school days. It is an exciting boxing tale that predates his one boxing novel, *Rodney Stone*. Robert Montgomery, a medical student strapped for funds (just like Doyle), took a job with an established practitioner in Sheffield. Like Doyle, he was the victim of a fouled-up bursary. Also like Doyle he was trying to cram five years' work into four; he was also overworked (certainly Doyle's situation with Dr. Hoare, a Dr. Elliot in Shropshire, and a Dr. Richardson in Sheffield). To make ends meet, the husky young student agrees to fight a 40-year-old local dreadnaught (he almost has to take on the wife, too)—the great Croxley Master.

Doyle did not earn any tuition with his fists, but he undoubtedly was a spectator on occasion of two popular student diversions: billiards and boxing on Lothian Street. As ship's doctor on the whaling ship *Hope*, he put on the gloves more than once, and like young Montgomery his fists won him many admirers. Doyle's account of the young student's relation-

ship with a Dr. Oldacre matches his account of his own three-month stint
with Dr. Richardson in Sheffield, Croxley being the "slum" or steel work-
ers' section of the city. Both young men found little time for themselves,
rolled countless pills, lived on skimpy diets, prided themselves on their
cricket game, and removed themselves from the situation as soon as possible.

The longest and best account of the young medical student, however,
certainly is found in *The Stark Munro Letters,* one of Doyle's comic mas-
terpieces. Although all of the action takes place right after graduation,
we constantly see the influence of Edinburgh's medical school.

John Stark Munro (probably named after three generations of Munros
on the Edinburgh medical faculty) is a thinly veiled disguise for A. Conan
Doyle. He writes to an American friend, mostly about his bizarre adventures with one Cullingworth, really George Budd, Jr., a classmate in
medical school. In real life both young men decided early on to do research
and write articles for the medical journals. Undoubtedly their university
stressed publishing far more than the other British medical schools. The
Edinburgh faculty filled the learned medical journals of the day. In the
Gaslight Publication's edition of *The Stark Munro Letters,* Dr. Frederick
Kittle's superb "Afterword" states that one of the strangest experiments
in the novel may be based on earlier research. In the very first letter the
two young men are seen dissecting and frying a diseased liver that Budd
stole from a pathology laboratory. They are trying to extract a waxy
substance, thus the frying technique. Dr. Kittle's belief is that Doyle is
recalling articles about amyloid degeneration written by Budd in 1879 and
1880 while he was still a student. Ironically, *On Diseases of the Liver* was
one of the best known works of Budd's father, also a doctor.

Doyle always averred that the Spartan life of the Edinburgh student,
the demands of the curriculum and faculty served him in good stead. It
was sink or swim. While many of the esteemed faculty members would
have little to do with the students, they served as models—in scholarship,
in internal discipline, and as conscientious (if not always competent)
medical men. Both Cullingworth and Stark Munro know how to make
do; they have seen poverty staring them in the mirror. They are undaunted.
Cullingworth is like a tank—nothing will stop him. More cynical than
Doyle about their training, young Budd determines he'll succeed even if
he has to blink at ethics; Doyle, ever the man of integrity, departs from
his friend after several petty squabbles, and decides to leave it up to fate
and his medical training to succeed.

One of the anecdotes that Joe Bell passed on to a *Strand* reporter
concerned a woman with an ulcerous cheek. None of his students (Doyle
among them) had the faintest idea of what was wrong. Like Mammy
Yokum, she smoked a short cutty (clay) pipe. One of Munro's first op-
erations is performed on a man with an ulcerous cheek. The cheek was

aggravated by the man's smoking a short cutty pipe most of his life. Young Stark Munro immediately made the diagnosis and began cutting. In one of many subtle one-liners in the novel, Doyle says that the man, when fully healed, "has just been in to tell me that he has bought a box full of churchwardens."

Finally, Doyle, who said in his famous comment that his detective must put into operation the method of "my old mentor Joe Bell," seems to have put the method into operation himself many years after he left the university. In a monograph published in 1960 by Bill Smith, of the *Copper Beeches,* an American doctor, Harold Gordon, a graduate of Edinburgh, recalls the occasion:

> I do recall very vividly two episodes of Sir Arthur's visit at Edinburgh during my student days. . . .I can't remember for sure which year he addressed us and later conducted a clinic. I believe it was 1912 or 1913. . . .
>
> He conducted ward rounds and I was one of the lucky ones to be present, because I was a dresser for Sir Robert Phillip whose specialty was tuberculosis and diseases of the chest. As we trooped behind Sir Arthur into one of the large public wards, he stopped suddenly just inside the entrance, sniffed a few times and exclaimed, "You have a typhoid patient here—I can smell it!" You may imagine our surprise as he stood still a moment or two longer, then walked over to one of the beds (it had a screen about it) and said to the patient, "Is your headache better now?" Sure enough, the man, pale but gaunt and still feverish, nodded his head as he said, "Aye, but thairsty the noo." The patient was still under observation, diagnosis uncertain. But with that as a lead, he was soon noticed to have a few ill-defined "typhoid spots" on his back and was later proven to have enteric fever. I suppose Sir Arthur remembered his experiences during the Boer War and arrived at the correct diagnosis by experience as well as by intuition. (I wish I could recall how the patient fared.)
>
> A little later, we stopped at the crib of a young baby, around two or two-and-a-half years old. The child's mother was watching at the lad's side. Almost without hesitation, Sir Arthur turned to the mother and said, kindly but with authority, "You must stop painting the child's crib." Sure enough, the child was in with lead poisoning. We were aware of the diagnosis, but asked how he had arrived at the right conclusion so quickly. He smiled as he answered, "The child looked pale but well-fed. He was listless and his wrist dropped as he tried to hold a toy. The mother was neatly dressed but she had specks of white paint on the fingers of her right hand. Children like to sharpen their teeth on the rails of a crib—so lead poisoning seemed a likely diagnosis!"

Little ever escaped the young medical student turned writer. The instructors, the standards, the classes, and the surroundings etched themselves in his mind. It is no accident that his famous detective forever played with test tubes in a corner of the room; Crum Brown and Joe Bell also experimented in their laboratories. Finally, when he got around to writing that first story with the new detective, the first victim is found at 3 Lauriston Gardens. When Conan Doyle was a junior in medical school, the Royal Infirmary of Edinburgh was moved to the Lauriston site. His memory was phenomenal; his own inner chemistry perked on all burners.

Bibliography

Bell, Joseph. "James Syme, Surgeon," in *Famous Edinburgh Students*. W. Scott Stevenson, Ed.; Edinburgh: T. N. Foulis, 1916.

Carr, John Dickson. *The Life of Sir Arthur Conan Doyle*. New York: Harper & Bros., 1949.

Chiene, John. "Looking Back: 1907–1860," in *Edinburgh Medical Journal XXII*, New Series, November 1907, 410–23.

Doyle, Arthur Conan. *The Annotated Sherlock Holmes*. William S. Baring-Gould, Ed.; 2 vols.; New York: Clarkson N. Potter, 1967.

Doyle, Arthur Conan. *Beyond the City*. Bloomington, IN: Gaslight Publications, 1982.

Doyle, Arthur Conan. *The Firm of Girdlestone*. London: John Murray, 1923.

Doyle, Arthur Conan. *The Green Flag and Other Stories of War and Sport*. London: Smith, Elder & Co., 1900.

Doyle, Arthur Conan. *Memories and Adventures*. Boston: Little Brown & Co., 1924.

Doyle, Arthur Conan. *Round the Red Lamp Stories*. London: Methuen & Co., 1894.

Doyle, Arthur Conan. *The Stark Munro Letters*. Bloomington, IN: Gaslight Publications, 1982.

Gordon, Harold. "Some Recollections of Sir Arthur Conan Doyle," edited, annotated, and expanded by Bill Smith in a monograph presented to the Philadelphia Sons of the Copper Beeches, 1960.

Gunn, Clement Bryce. *Leaves from the Life of a Country Doctor*. Rutherford Crockett, Ed.; Edinburgh: Moray Press, 1935.

Jex-Blake, Sophia. *Medical Women*. Edinburgh: W. Oliphant, 1886.

Mutrux, Henri. *Sherlock Holmes: Roi des Tricheurs*. Paris: Pensée Universelle, 1977.

Symons, Julian. *Portrait of an Artist: Conan Doyle.* London: Whizzard Press/Andre Deutsch, 1979.

Turner, A. Logan. *Story of a Great Hospital: The Royal Infirmary of Edinburgh, 1729–1929.* Edinburgh: Oliver & Boyd, 1937.

Wallace, Irving. *The Fabulous Originals.* New York: Alfred A. Knopf, 1955.

Strong Poison
Chemistry in the Works
of Dorothy L. Sayers

Natalie Foster

When Alistair Cooke introduced the first BBC dramatization of a Lord Peter Wimsey mystery to American audiences, he stated that there were two classes of mystery readers: those who have never read Dorothy L. Sayers, and those who have never read anybody else.[1] Actually Sayers made an indelible mark on three literary spheres. A discrete sign placed on her former home in Witham, Essex, by the local Preservation Society testified laconically to this by identifying her as a novelist, theologian, and Dante scholar.[2] But part of the special fascination of her detective works stems from the vast amount of painstaking research that she undertook on topics related to the characters and situations she created. Such research also was characteristic of her religious plays and essays and her translations of medieval literature. Her use of chemistry in several novels and short stories bears witness to her desire to be precise and correct in what she wrote and to do whatever may be necessary "to make a good book."[3]

Dorothy L. Sayers was born in Oxford in 1893, the only child of the Reverend Henry Sayers and his wife Helen.[4] Although she was not trained in the sciences (she graduated from Somerville in 1915 with Class I honors in French),[5] her mysteries are laced with fact and opinion on the science of the 1920s and 1930s.

Muscarine in Mushrooms

Decidedly the most chemically oriented of her works was a novel written in collaboration with Robert Eustace in 1930 entitled *The Documents in the Case*. Robert Eustace was the pen name of Eustace Robert Barton, a physician (member of the staff at the Gloucester Mental Hospital), and contemporary of Sir Arthur Conan Doyle, who often lent his advice to authors seeking to incorporate scientific information into their novels.[6,7]

Briefly, *The Documents in the Case* deals with the death of an authority on wild edible mushrooms after he eats a stew concocted from the fungi he had gathered in the woods. The coroner's inquest offers a verdict of accidental death, concluding that the gentleman, George Harrison by name, mistook the mushroom *Amanita muscaria* (the poisonous deadly amanita or fly agaric) for *Amanita rubescens* (a nonpoisonous, edible variety). The mistake resulted in his accidental death by muscarine poisoning.

Pharmacologically and forensically, this was an excellent choice for a poisoning case, because 90% of all fatalities from mushroom poisoning are caused by the ingestion of a poisonous member of the Amanita family.[8] Also, as early as the turn of the century, the physiological responses to muscarine had been well characterized. Muscarine, in fact, is known as the "foundation stone of modern pharmacology" because its preparations were the first chemicals to reproduce faithfully some of the natural responses to stimulation of the parasympathetic nervous system,[9] thereby giving rise to terms like "muscarine response" or "muscarinic action," which are still in use.

The progression of the novel, however, turns on the disbelief on the part of Harrison's son that his father could have made such a pedestrian error. No one can imagine how the man could possibly have been murdered given the circumstances of the novel until the young protagonist, John Munting, hears a lesson in current chemistry at a cocktail party. At this gathering is a chemist named Waters, "the coming man in chemistry at Oxford,"[10] who responds to an initially philosophical question on the weighty topic of "what is life" in a uniquely chemical fashion:

> At present—chemically speaking—the nearest definition I can produce is that it is a kind of bias—a lop-sidedness, so to speak. Possibly that accounts for its oddness. . .
>
> . . .up to the present, it is only living substance that has found the trick of transforming a symmetric, optically inactive* compound into a single, asymmetric, optically active compound. At the moment that Life appeared on this planet, something happened to the molecular structure of things. They got a twist, which nobody has ever succeeded in reproducing mechanically—at least not without an exercise of deliberate selective intelligence, which is also, as I suppose you'll allow, a manifestation of Life.
>
> . . .If you pass [a light] ray through a crystal of Icelandic spar [*see* box], the vibrations are all brought into one plane, like a flat ribbon. That is

---

*The Avon Books edition of this novel has the word "active" at this point. This may be a typographical error. "Inactive" must be the intended word to describe the behavior of a *symmetric* compound with polarized light.

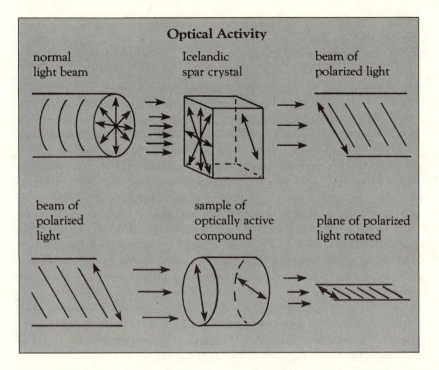

Optical Activity

normal light beam   Icelandic spar crystal   beam of polarized light

beam of polarized light   sample of optically active compound   plane of polarized light rotated

what is called a beam of polarized light. . . If you pass this polarized light through a substance whose molecular structure is symmetric, nothing happens to it; the substance is optically inactive. But if you pass it through, say, a solution of cane sugar, the beam of polarized light will be twisted, and you will get a spiral effect, like twisting a strip of paper either to the right or to the left. The cane sugar is optically active. And why? Because its molecular structure is asymmetric. The crystals of sugar are not fully developed. There is an irregularity on one side, and the crystal and its mirror image are reversed, like my right hand and my left." He laid the palm of the right hand on the back of the left to show his meaning.

. . .we can produce in the laboratory, by synthesis from inorganic substances, other substances which were at one time thought to be only the products of living tissues. . . But what is the difference between our process and that of nature? . . . The substance produced by synthesis always appears in what is called a racemic form. It consists of two sets of substances—one set having its asymmetry right-handed and the other left-handed, so that the product as a whole behaves like an inorganic, symmetric compound; that is, its two asymmetries cancel one another out, and the product is optically inactive and has no power to rotate the beam of polarized light.[11]

Waters concludes by explaining that the right- and left-handed forms may be separated by the exercise of our living intelligence but that the task is by no means trivial and indeed proves to be quite difficult.

Despite a few phrases of questionable meaning ("crystals. . .not fully developed," "inorganic, symmetric compound") and the general feeling that this type of optical activity is the property of a crystal and not molecular in origin, most chemists would have to agree that the author has provided a lucid and essentially correct description of the phenomenon of optical activity. In the course of this discussion during the party, Munting realizes with a sinking feeling that Waters has told him how the crime could have been committed with synthetic muscarine. Munting knows that his colleague Philip Lathom is the likely perpetrator of the crime, and he feels torn between doing his duty and loyalty to his friend. Munting questions Waters in private about the possibility and Waters suggests that they go immediately to the forensic chemist in charge of the investigation and request that he check the samples of stew for their optical activity. The muscarine in the stew proves optically inactive—ergo, it is synthetic; therefore, Harrison was murdered. Munting observes at the end of the novel that he would hate to be in Lathom's place and have been tripped up by a miserable asymmetric molecule. (Incidentally, this is not an uncommon lament among beginning organic chemistry students facing their first examination questions on stereochemistry.)

From a scientific vantage point, one might wonder just how accurate and up-to-date Sayers' chemistry was throughout the novel. Despite the precise pharmacological characterization of muscarine, the substance has proven extraordinarily elusive and only recently (1957) have its isolation and the determination of its structure been achieved.[9] The forensic chemist in the novel gives a rather good, albeit brief, summary of research on muscarine up to 1930:

> The isolation of muscarine itself in a pure state from the fungus would be a chemical experiment of considerable difficulty, and has, so far as I know, been accomplished only by two men, Harnack and Nothnagel; their results have not, I believe, received confirmation as yet. Choline aurichloride and muscarine aurichloride have been obtained by Harnack from fractionation of extracts of the fungus, and, more recently, King obtained muscarine chloride from the same source.[12]

In point of fact, Harnack[13] in 1875 isolated muscarine aurichloride from a syrupy extract of Amanita and suggested the formula for the active part of the molecule (Structure 1). In 1893 and 1894, Nothnagel[14] repeated and verified Harnack's synthetic substantiation of this formula by oxidizing choline with nitric acid and thereby creating what was called for several years "synthetic muscarine." Doubt of the correctness of this formula persisted until 1914 when Ewins[15] proved that the reaction produced the nitrous ester of choline (Structure 2) and not a material analogous to muscarine itself.

In 1922 a drastic revision in the extraction procedure enabled King[16] to isolate the purest sample of natural muscarine up to that time, but King's work and the formula he proposed ($C_8H_{18}O_2N$; this is a molecular formula and has no assigned structure associated with it) were never substantiated. A further period of inactivity followed until 1931 (recall that the novel was published in 1930) when Kögl and coworkers[17] isolated what they believed to be a pure muscarine salt and determined the material to be optically active. Not until Eugster and Wasser (1954–1956)[18] utilized partition chromatography on cellulose columns was analytically pure muscarine isolated and its optical activity verified. The final structural determination was made in 1957 by Jellinek[19] using x-ray crystallography (Structure 3), and the total synthesis of muscarine was accomplished shortly thereafter.[20]

With this background, let us return to the novel and see how well Sayers and Eustace did. The formula cited for muscarine by an eager university student taking Munting and Lathom on a tour of his chemistry building was $C_5H_{15}NO_3$ (analogous to the hydroxide salt of Structure 1), which was known by 1914 to be incorrect. The student also mentions the preparation as the one known to yield the choline nitrous ester. In response to a question by Lathom regarding the synthetic muscarine that the student has just proudly pointed out to them on the laboratory shelf, the student replies:

> (It's made of) inorganic stuff, you know—all artificial. . . You start
> with artificial choline. . . you can make it by heating ethylene oxide with

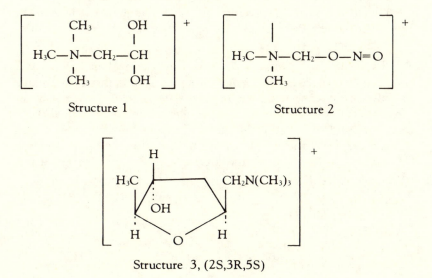

Structure 1   Structure 2

Structure 3, (2S,3R,5S)

triethylamine [*sic*]. That gives you your choline. Then you oxidize it with dilute nitric acid—the stuff you etch with, you know. Result, muscarine. Pretty, isn't it?

[Lathom:] And if you analyze it again chemically, could you tell the difference between that and the real stuff?

[Student:] Of course not. It *IS* the real stuff.[21]

The eager young chemist apparently does not know anything about optical activity in natural products for all his glib synthetic knowledge. At any rate, Lathom was not only tripped up by a miserable asymmetric molecule; he was ruefully misinformed by a second-year graduate student.

In all fairness to Sayers and Eustace regarding the formula, the reference they used was probably J. Dixon Mann's *Forensic Medicine and Toxicology* (Sayers actually has Lord Peter Wimsey consulting a volume of Mann in *Strong Poison*),[22] and the 1922 edition of this treatise cites the formula proposed by Harnack for muscarine.[23] Harnack's structure is incapable of optical activity because it lacks a chiral carbon. * By her own admission, Sayers later realized that the plot was fine in this novel but the toadstool was not.[6] After the novel was published, a chemist wrote to her that her general theory was sound but that natural muscarine was an exception to the basic idea in that it did not rotate the plane of polarized light. He was correct, based on the knowledge of 1930. In the last analysis, however, both her theory and the crime were totally and unambiguously sound. The chemistry of her day—which later proved inaccurate—would not allow her to poison her victim and catch the murderer the way she chose, but modern day science had vindicated her idea of trapping this particular criminal with a polariscope.

Two other small items are worthy of note: the synthesis of muscarine cited in the novel has an error (trimethylamine, not triethyl), but whether the mistake is a conceptual one or merely typographical can not be ascertained. Another technical error, originally pointed out by Harold Hart,[6] comes in the description of the scene in the forensic laboratory during the measurement of the optical activity of the evidence with a polariscope: "He snapped off the lights, and we were left with only the sodium flame. In that green, sick glare. . ."[24] Of course, a sodium flame is bright yellow-orange, but sick green does seem to set the tone more appropriately for a grisly investigation.

---

*A chiral carbon is a carbon atom with four different groups attached to it; in all but a few unique organic structures a chiral carbon is necessary for a compound to display optical activity.

Thyroxine in Wafers

A second illustration of Sayers' astute use of chemistry, this time principally medicinal chemistry, comes in a short story published in 1933, "The Incredible Elopement of Lord Peter Wimsey." In this adventure she has the opportunity to apply the medicinal knowledge of "glands" that she hinted on in several earlier works. "Glands" had apparently been the rage in England for several decades:

> Glands, my child, glands are the thing.[25]

> He's founding a new clinic to make everybody good by glands. . . It's the science of the future as they say in the press. There really isn't any doubt about that. It puts biology in quite a new light.[26]

> When in doubt, give thyroid.[27]

The story takes place in a remote Basque village where the local peasants refer to a young American lady resident as bewitched. Wimsey correctly diagnoses the source of the medical problem with this former beauty. She suffers from myxedema, a thyroid deficiency disease, the extreme manifestations of which Sayers describes with harsh clinical precision:

> There was nothing to be seen, but a noise had begun; a kind of low, animal muttering, extremely disagreeable to listen to. It was not made by a dog or a cat, he felt sure. It was a sucking, slobbering sound that affected him in a curiously sickening way. It ended in a series of little grunts and squeals, and then there was silence. . . Something shuffled and whimpered among the cushions. . . . It was dressed in a rich gown of gold satin and lace that hung rucked and crumpled upon the thick and slouching body. The face was white and puffy, the eyes vacant, the mouth drooled open, with little trickles of saliva running from the loose corners. A dry fringe of rusty hair clung to the half-bald scalp, like the dead wisps on the head of a mummy. . . (The inert hand) was clammy and coarse to the touch and made no attempt to return the pressure. . .[28]

The crime involves the lady's own jealous and vengeful husband who intentionally withholds from her the corrective medicine she needs to counter her disorder and thereby reduces her to cretinous imbecility. Wimsey restores the lady to health by smuggling to her a weekly consignment of thyroxine-containing wafers while the evil husband is abroad.

Research on the thyroid gland and the total synthesis of the active principle thyroxine had been going on for almost thirty years and culminated in some important scientific discoveries just a few years prior to the publication of this story. That much of the early work on thyroxine was done by British scientists may help to explain some of Sayers' interest and knowledge of the subject.

In 1891 Murray[29] observed that when he grafted sheep thyroid to the neck of a myxedemic woman the condition previously thought to be incurable improved. The improvement was so sudden that Murray hypothecated that some substance contained in the gland was being slowly secreted into the body of the woman. This substance must be responsible for the improvement rather than any action of the transplanted gland itself. Fox and MacKenzie,[30] working independently, both reported successful treatment of myxedema by prescribing the oral administration of preparations of the thyroid gland "to be taken lightly fried with currant jelly" or "minced and moistened with brandy."

The chemical analysis of thyroxine was advanced by Eugene Baumann at the University of Freiburg in 1895 upon his discovery of the presence of iodine in the active principle of the gland extract.[31] Further work on the isolation and characterization of the hormone was delayed by Baumann's untimely death at the age of 49 in 1896 due to a chronic heart condition. E. C. Kendall, working at the Mayo Clinic with support from Parke–Davis, proposed a structure for thyroxine in 1919 (Structure 4) that was later shown to be erroneous. C. R. Harington, a young British scientist, had first become involved in the thyroxine problem in 1922 while on a post-doctoral year with H. D. Dakin at the Rockefeller Institute in New York. Upon his return to Britain, Harington continued his studies while lecturing at University College Hospital Medical School in London where he ultimately established in 1927 the correct structure of thyroxine (Structure 5) by total synthesis.[32] Interestingly, Harington's mentor, Dakin, had simultaneously assigned the same structure to thyroxine but withdrew the paper announcing his result when he learned of Harington's work and thereby allowed Harington the credit of a unique discovery.[33] As a result of this work, Harington was named the head of the British National Institute for Medical Research and was later knighted. The climate of

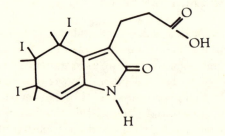

Structure 4

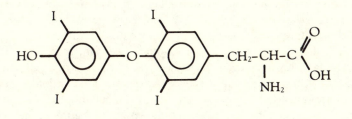

Structure 5

research and discovery in Britain during this period may have led Sayers to the unique premise for this story.

Regarding the plot once again, Wimsey's analysis of the situation and his correct diagnosis are models of deductive reasoning laced with medical awareness:

> . . . To begin with, there was this extraordinary kind of decay or imbecility settlin' in on a girl in her twenties. . . And then there was this tale of the conditions clearin' up regularly once a year or so—not like any ordinary brain trouble. Looked as if it was being controlled by somebody. . . Mrs. Wetherall had been under her husband's medical eye from the beginning. . . Then there were Wetherall's well-known researchers, and the fact that he kept in touch with a chemist in London. . . Alice Wetherall is one of those unfortunate people who suffer from congenital thyroid deficiency. You know the thyroid gland in your throat—the one that stokes the engine and keeps the old brain going. In some people the thing doesn't work properly, and they turn out cretinous imbeciles. Their bodies don't grow and their minds don't work. But feed 'em the stuff, and they come out absolutely all right— cheery and handsome and intelligent and lively as crickets. Only, don't you see, you have to *keep* feeding it to 'em, otherwise they just go back to an imbecile condition.[34]

Arsenic in Hair and Fingernails

A third selection that demonstrates Sayers' up-to-date knowledge of science is *Strong Poison* (1930) in which Harriet Vane, the future wife of Peter Wimsey, stands accused of murdering her erstwhile lover Philip Boyle by poisoning him with arsenic. It is interesting to speculate that Sayers may have developed the ideas behind "The Incredible Elopement. . ." and *Strong Poison* simultaneously. Available treatises on thyroid therapy contained considerable speculation on the relationship between the therapeutic effect of arsenic and its action on the thyroid.[35] Around the turn of the century arsenic was one of the standard drugs used in the treatment of Graves' disease (goitrous thyroid gland, or hyperthyroidism) and any research into the action of either thyroxine or arsenic was bound to uncover facts about the other.

Arsenic has long been a favorite murder weapon, both literarily and in actual fact, but here again Sayers weaves some scientific acumen into the fabric of her tale. She not only mentions the folk wisdom about the Styrian peasants who were known arsenic eaters, but also she refers directly (again by citing Dixon Mann's text) to the series of experiments by Coletta in 1906[36] that established the phenomenon of localized immunity to arsenic as a documented fact.

> . . . about this arsenic. As you know, it's not good for people in a general way, but there are some people—those tiresome peasants in Styria one hears so much about—who are supposed to eat it for fun. It improves their wind, so they say, and clears their complexions and makes their hair sleek, and they give it to their horses for the same reason; bar the complexion, that is, because a horse hasn't much complexion. . . it's well known that some people do take and manage to put away large dollops after a bit of practice— enough to kill an ordinary person. . .

> . . . some fellow—I've forgotten his name, but it's all in Dixon Mann— wondered how the dodge was worked, and he got going on some dogs and things and he dosed 'em and killed a lot of 'em I dare say, and in the end he found that whereas liquid arsenic was dealt with by the kidneys and was uncommonly bad for the system, solid arsenic could be given day by day, a little bigger dose each time, so that in time the doings—what an old lady I knew in Norfolk called the "tubes"—got used to it and could push it along without taking any notice of it, so to speak. I read a book somewhere that it was all done by leucocytes—those jolly little white corpuscles, don't you know—which sort of got round the stuff and bustled it along so that it couldn't do any harm. At all events, the point is that if you go on taking solid arsenic for a good long time—say a year or so—you establish a what-not, an immunity, and can take six or seven grains at a time without so much as a touch of indijaggers. . .

> Apparently these beastly Styrian peasants do it that way, and they're very careful not to drink for two hours or thereabouts after taking it, for fear it should all get washed into the kidneys and turn poisonous on 'em.

> . . . as I say, you have a nice clear complexion—except that I notice the arsenic has pigmented the skin here and there (it does that sometimes), and you've got the sleek hair. . . and I notice you were careful not to drink at dinner. . . Then we got hold of some bits of your hair and nails, and lo and behold, they were bung-full of arsenic.[37]

This explanation by Lord Peter to the suspected murderer is a very good summary of the information known at that time about the physiological effects of arsenic and the technique for building up immunity.[38] Regarding the analysis of the samples of fingernails and hair from the suspected murderer, Wimsey's manservant Bunter performs a credible Marsh's test for the presence of arsenic and even remembers to test his reagents for arsenic content before commencing the actual evaluation of the forensic samples:

> ". . . The distilled water was already bubbling gently in the flask. . . You will perceive that the apparatus is free from all contamination."
>
> "I see nothing at all."
>
> "That, as Sherlock Holmes would say, is what you may expect to see when nothing is there."
>
> . . . And presently, definitely, magically, a thin silver stain began to form in the tube where the flame impinged upon it. Second by second it spread and darkened to a deep brownish-black ring with a shining metallic center. . .
>
> "It's either arsenic or antimony. . . The addition of a small amount of solute of chlorinated lime should decide the question. . ."
>
> The stain dissolved out and vanished under the bleaching solution.
>
> "Then it's arsenic."[39]

Even in the use of arsenic, which is certainly nothing new in the tomes of literary crime, Sayers buttresses her story with factual knowledge of the science behind the phenomenon. Other sections of the novel include medically accurate descriptions of the physiological response to a lethal dose of arsenic, biodistribution and accumulation of arsenic in living tissues, and even a concise reference to the packaging laws in Britain governing the sale of arsenic at the time of the murder.

Lest one think that these are the only examples of science in Dorothy L. Sayers' mysteries, some mention is due Eric Loder who, in "The Abominable History of the Man with Copper Fingers," electroplates his victims with a suitable solution of cyanide ion and copper sulfate, Dr. Walter Penberthy in *The Unpleasantness at the Bellona Club*, who gently overdoses an elderly gentleman with his own trusted heart medicine, George Harrison in *The Documents in the Case*, who reveled in discussions of Mr. Einstein and his theories, and John Munting in the same novel, who sheds some light on causality in literature through the second law of thermodynamics. Factual science and clinically accurate medicinal chemistry are mainstays in the detective fiction of Dorothy L. Sayers.

References

[1] Cooke, Alistair, private communication, January 14, 1982.
[2] Winn, Dilys. *Murderess Ink: The Better Half of the Mystery*. New York: Workman Publishing, 1979; p. 56.

[3] Hitchman, Janet. *Such a Strange Lady.* New York: Harper and Row, 1975; pp. 75, 155; Hone, Ralph E. *Dorothy L. Sayers: A Literary Biography.* Kent, OH: Kent State Univ., 1979; pp. 75, 77.

[4] Ibid.; p. 5.

[5] Ibid.; p. 21.

[6] Hart, Harold. *J. Chem. Ed.* **1975**, *52*, 444.

[7] Horn, Ralph E. op. cit.; p. 196.

[8] Goodman, L. S.; Gilman, A. *The Pharmacological Basis of Therapeutics.* New York: Macmillan Co., 1965; p. 475.

[9] Wilkinson, S. *Quart. Rev.* **1961**, *15*, 153.

[10] Sayers, D. L. with Eustace, Robert. *The Documents in the Case.* New York: Avon Books, 1968; p. 201.

[11] Ibid.; pp. 206–7.

[12] Ibid.; p. 178.

[13] Harnack, H. *Arch. Exp. Pathol. Pharmakol.* **1875**, *4*, 168.

[14] Nothnagel, G. *Ber. Dtsch. Chem. Ges.* **1875**, *26*, 801.

[15] Ewins, A. J. *Biochem. J.* **1914**, *8*, 209.

[16] King, H. J. **1922**, *121*, 1843.

[17] Kögl, F.; Duisberg, H.; Exerleben, H. *Ann.* **1931**, *489*, 156.

[18] Eugster, C. H.; Waser, P. G. *Experientia* **1954**, *10*, 298; Eugster, C. H. *Helv. Chim. Acta* **1956**, *39*, 1002; Waser, P. G. *Experientia* **1955**, *11*, 452.

[19] Jellinek, F. *Acta Cryst.* **1957**, *10*, 277.

[20] Wilkinson, S. op. cit. and references cited therein.

[21] Sayers, D. L. *The Documents in the Case,* op. cit.; pp. 195–96.

[22] Sayers, D. L. *Strong Poison.* New York: Avon Books, 1967; p. 187.

[23] Mann, J. Dixon. *Forensic Medicine and Toxicology,* 6th ed.; London: Charles Griffin & Co., Ltd., 1922; p. 542.

[24] Sayers, D. L. *The Documents in the Case.* p. 216.

[25] Ibid.; p. 25.

[26] Sayers, D. L. *The Unpleasantness at the Bellona Club.* New York: Avon Books, 1963; pp. 122–23.

[27] Waller, H. E. *Theory and Practice of Thyroid Therapy.* London: John Bale, Sons and Danielsson, Ltd., 1912; preface.

[28] Sayers, D. L. "The Incredible Elopement of Lord Peter Wimsey," in *Lord Peter.* compiled by James Sandoe. New York: Avon Books, 1972; pp. 308–10.

[29] Murray, G. R. *Br. Med. J.* Oct. 10, 1891; 796.

[30] As quoted in Harington, C. R. *The Thyroid Gland: Its Chemistry and Physiology.* London: Oxford Univ., 1933; p. 14.

[31] Wolff, Manfred E., Ed. *Burger's Medicinal Chemistry,* 4th ed., Part III; New York: John Wiley and Sons, 1981; p. 105.

[32] Harington, C. R.; Barger, G. *Biochem. J.,* Vol 1927, 169.

[33] Wolff, Manfred E. op. cit.; p. 107.

[34] Sayers, D. L. "The Incredible Elopement of Lord Peter Wimsey"; pp. 325–26.

[35] Waller, H. E. op. cit.; p. 118.

[36] Cloetta, S. *Arch. Exp. Pathol.* 1906; as cited in Mann, J. Dixon. *Forensic Medicine and Toxicology.* op. cit.; p. 370.

[37] Sayers, D. L. *Strong Poison,* op. cit.; pp. 187–88.

[38] As described in Mann, J. Dixon. op. cit.; pp. 307*ff.*

[39] Sayers, D. L. *Strong Poison,* op. cit.; pp. 177–79.

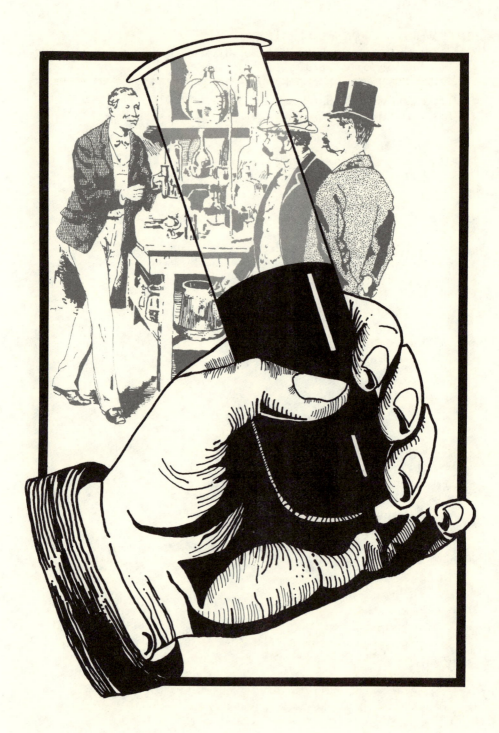

3

A Study in Scarlet
Blood Identification in 1875

Samuel M. Gerber

In the first of the Sherlock Holmes stories, "A Study in Scarlet," Watson and Holmes meet in 1881. On that occasion, Holmes announces the discovery of a new test for blood and extols its specificity, sensitivity, and simplicity. The test was indeed simple: "a drop of blood, a liter of water [1 part blood in about 20,000–30,000 parts water], a few white crystals, a drop of a transparent fluid—a dull mahogany color to a brownish dust." What was the chemical nature of the test? Let us first review how human blood was identified in 1875.

Testing for Blood in 1875

The first identification was visual. The investigator simply looked at the color, the gleam of dried blood, the stiffness of the fabric where blood had dried, and other such factors.

Chemical analysis was in use in 1875 for blood identification. These tests were *presumptive*, that is, if the test is positive, the investigator should continue with more specific tests.

The first chemical visual test is simple: a solution of water and the suspected sample is prepared. If the solution contains blood, it does not change color when dilute ammonia is added; it turns brown, however, when concentrated ammonia is added. Heating causes it to coagulate.

The *guaiacum test* was the important test for blood in 1875. The test is attributed to John Day of Australia (about 1867–1869) with other claims by Schonbein and Ven Deen. Guaiacum, which is a tree resin, is added to a water solution of suspected blood; then hydrogen peroxide is added. If the material is blood, a blue color is formed. Adding alcohol gives a sapphire blue.

The sensitivity has been variously reported, from 1 in 2000 to 1 in 100,000 parts blood in water. I suspect that the sensitivity depended on

0784/83/0031$06.00/0
© American Chemical Society

the purity of the guaiacum resin. The precise composition of the resin is not known; however, guaiacol (*see* structure) is obtained from the resin. Compounds similar to guaiacol form the dye shown as aurine by oxidation. A more complex material would give deeper shades and would approach a red or a reddish brown in appearance.

Microscopic examination of blood was of limited use for two reasons. First, microscopic tests could distinguish the blood of birds, reptiles, fish, and mammals, but not between humans and other mammals. Second, dried blood could not be used.

In the *hematin* test, now referred to as the Teichmann test, blood is mixed with salt crystals and glacial (concentrated) acetic acid and heated. The blood cells are evaporated, and rhombic crystals of hemin are formed. Hemin is the product of chloride and heme, a blood constituent responsible for the oxygen-carrying capacity of blood. Thus, the hematin test proves only that the material is mammalian blood.

The Sherlock Holmes Test

Considering the existing state of the art just prior to 1881, what can we say about the Sherlock Holmes test for blood identification? To cause blood to change its color from red to mahogany as Holmes describes, we need an acid to increase its oxidation rate, and a material to be oxidized. Some possibilities for the "few white crystals" and the "drop of transparent fluid" that Holmes could have used to produce the dark color are shown in Table I. It is difficult to know whether such systems are sensitive to 1 part blood per 1,000,000 parts water and whether they are specific to human blood. The sensitivity probably would be similar to that of the guaiacum test.

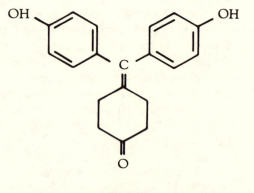

Guaiacol

Aurine

Table I. The Sherlock Holmes Test

Colorless Crystals (Oxidant)	Clear Liquid (Acid)	Color Former (Transparent Fluid)
Sodium peroxide	Acetic acid	Nitrosophenol + dimethylaniline
Sodium perborate	Propionic acid	Nitroso-α-naphthol

Table II. Some Chemical Tests for Blood Identification Since 1875

Color Former	Color Formed	Sensitivity	Discoverer
Benzidine	Blue	1 part/million	Adler & Adler (1904)
Malachite green (Leuco)	Green	—	—
Phenolphthalein	Pink	1 part/6 million	Kastle-Meyer
Luminal (2-amino-phthalhydrazide hydrochloride)	Luminesces	—	—
Haemochromogen sodium hydroxide, pyridine, glucose	Pink needles	—	Takayama

Testing for Blood Since 1875

Most tests for blood are based on the same principle: Peroxidase, an enzyme in blood, acts as a catalyst for the oxidation of a chemical material that forms a characteristic color (see Table II).

Benzidine was the chemical of choice until recently (see structure). Benzidine was characterized in 1845 but not used in forensic medicine until 1904. The special reagent Holmes used could have been benzidine. Eventually, benzidine was clearly demonstrated to be a carcinogen, and so a replacement was sought, especially as a dyestuff intermediate. Dianisidine, tolidine, and dichlorobenzidine (see structures) have all been used. Tetramethylbenzidine is free of carcinogenicity, but it has not come into general use for various reasons. Phenolphthalein remains in active use. Antipyrine also has been used as a benzidine replacement.

Other chemical reagents listed in Table II are used for presumptive tests for blood; the tests of blood with the benzidine derivatives are acceptable but they are not definitive.

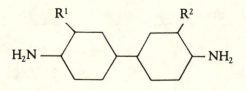

$R^1 = R^2 = H$ Benzidine

$R^1 = R^2 = CH_3O$ Dianisidine

$R^1 = R^2 = CH_3$ Tolidine

$R^1 = R^2 = Cl$ Dichlorobenzidine

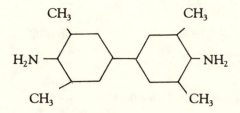

Tetramethylbenzidine

The major breakthrough in specifically identifying human blood came as the result of the work of Karl Landsteiner. He was a medical doctor, but he had a thorough training in chemistry. He discovered that human blood differed from person to person in the capacity of serum (the liquid portion) to cause the red cells to clump. That is, serum from patient 1 might clump red cells from patient 2 but not those of patient 3; serum from patient 2 might clump red cells from patients 1 and 3; and so on.

By 1902, Landsteiner had determined that human blood could be divided into four groups: A, B, AB, and O. Once this was known, it was a simple matter to show that transfusions between persons of the same group would have no ill effects, but transfusions between persons of different groups could cause the death of the recipient.

By 1910, it had also been determined that blood groups were inherited according to standard genetic laws, and thus they could be used as tools in paternity suits and other criminal investigations.

In 1927 Landsteiner and colleagues discovered the M, N, and MN blood groups, and in 1940, the Rh blood groups. In 1930, Landsteiner was awarded the Nobel Prize in medicine and physiology.

Chapters 7 and 8 discuss present-day sophistication in blood identification and characterization in more detail. Perhaps the story "A Study
in Scarlet" by Conan Doyle gave impetus to the development of improved
methods in blood identification.

References

[1] "A Study in Scarlet" from "The Complete Sherlock Holmes," Doyle, A. C.,
New York: Doubleday & Co., 2 vol., 1960.
[2] Graham, R. P. *J. Chem. Ed.* **1945**, 508.
[3] Taylor, A. S. "A Manual of Medical Jurisprudence," Philadelphia: Henry C.
Lea, 1873.
[4] Lucas, A. "Forensic Chemistry and Scientific Criminal Investigation," 4th ed.;
New York: E. Arnold and Co., 1946.
[5] Polson, C. J. "The Essentials of Forensic Medicine," London: Eng. Univ., 1955.
[6] "Glaister's Medical Jurisprudence and Toxicology," 13th ed., Edinburgh: Liv-
ingstone, 1973.
[7] Tedeschi, C. G.; Eckert, W. G.; Tedeschi, L. G. "Forensic Medicine," New
York: Saunders, 1977.

TO TODAY'S COURTROOM

Sherlock Holmes   Edmond Locard   magnifying glass   blood identification   soil analyses
fingerprinting   ballistics   document examination   mass spectrometer   scanning electron microscope   microspectrophotometer   computer   Sherlock Holmes   Edmond Locard   magnifying
glass   blood identification   soil analyses   fingerprinting   ballistics   document examination

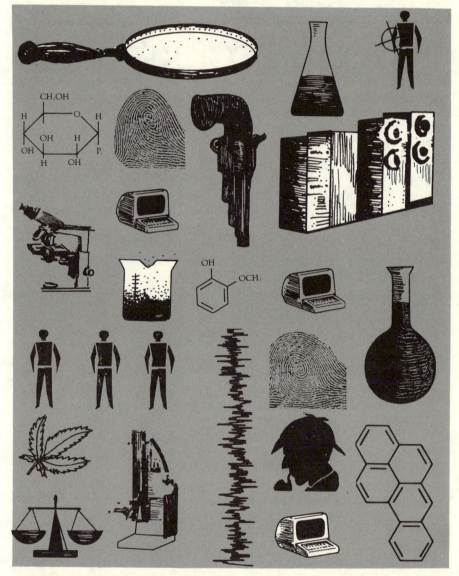

mass spectrometer   scanning electron microscope   microspectrophotometer   computer   Sherlock Holmes   Edmond Locard   magnifying glass   blood identification   soil analyses   fingerprinting   ballistics   document examination   mass spectrometer   scanning electron microscope
microspectrophotometer   computer   Sherlock Holmes   Edmond Locard   magnifying glass

4

Forensic Science
Winds of Change

Richard Saferstein

The idea of using science as an aid in criminal investigation was foreshadowed in the fictional works of Sir Arthur Conan Doyle at the turn of the century. However, Holmes' creator could scarcely imagine the resources that would eventually be applied to such endeavors. Currently in the United States there are nearly 250 public laboratories employing the services of about 3500 scientists in the scientific examination of crime scene evidence. These numbers do not even begin to reflect the installations and personnel involved in medico-legal investigations and the hundreds of private consultants and private laboratories engaged in forensic analytical services. The statistics for foreign forensic laboratories are likewise impressive. For example, England and Wales alone employ more than 600 scientists in nine government laboratories.

In succeeding chapters of this book, authors will describe a variety of analytical techniques that have come to be accepted as routine tools of the modern forensic scientist. However, before we begin our exploration of this subject it must be emphasized that forensic scientists, and in particular forensic chemists, should not be viewed as being engaged in an endeavor that is incompatible with other branches of analytical chemistry. The specimens forensic analysts must subject to examination are very different from what other scientists are normally asked to deal with, but their techniques and philosophies more often than not coincide. Few analytical chemists would feel uncomfortable or befuddled in the environs of a crime laboratory. In fact, one of the major impediments to continued progress in forensic science research has been a failure to merge forensic chemistry into the mainstream of analytical chemistry. This failure has inhibited communication between forensic scientists and their colleagues in the general analytical chemistry profession and has discouraged analytical chemists from applying the spinoffs of their work to forensic science problems. Publications such as this book can provide a firm basis for a

0784/83/0039$06.00/0
© American Chemical Society

mutual exchange of data and ultimately lead to a cross-fertilization of ideas among all scientists engaged directly or indirectly in forensic science-related research.

The Beginnings

Forensic scientists have been major beneficiaries of recent advances and improvements in analytical instrumentation. However, the luxury of access to sophisticated analytical equipment was one not always afforded the forensic science community. The famous French criminalist, Edmond Locard, inspired by the "Adventures of Sherlock Holmes" and the writings of other contemporaries, prevailed upon police authorities in Lyon to help him start what is generally regarded as the first crime laboratory. In 1910, with only a microscope and a spectroscope at his disposal, Locard started his laboratory in two rooms above the local courthouse. He became expert at characterizing dust and small particles, believing that any contact between two surfaces would result in a mutual exchange of trace evidence. Locard demonstrated the viability of his exchange principle in a number of famous and well-publicized cases. His successes ultimately laid the foundation for the ever-increasing role science was to play in criminal investigation.

Locard and his contemporaries prided themselves on their general knowledge and their abilities to apply a spectrum of skills to the diversity of items typically retrieved at crime sites. Few areas of knowledge escaped their attention or inquisitiveness. Even the indefatigable Sherlock Holmes was known to dabble in blood identification, soil analyses, fingerprinting, ballistics, and document examination.

Forensic Scientists Today

Today, we have had almost a complete turnabout in philosophy. A myriad of analytical instrumentation and the explosive expansion of scientific knowledge prevent all but a few gifted individuals from mastering the intricacies of the many fields of knowledge encompassing the practice of forensic science. As in many other fields of human endeavor, a high degree of specialization and teamwork is required to tackle the scientific examination of crime scene evidence. The operation and maintenance of complex equipment, such as the mass spectrometer or the scanning electron microscope, require talents and expertise far different from those of an analyst engaged in the characterization of bloodstains or synthetic fiber examinations. Often, interpretive skills must be forged through experience arising out of repetitive examinations. This is particularly true for examiners of hairs, toolmarks, bullets, and documents. Here, years of experience and the quantity of evidence examined are paramount criteria in judging the competence and expertise of the examiner. Other specialties require skills molded from a combination of theoretical and practical know-how. For example, genetically controlled protein polymorphisms detected in bloodstains exhibit electrophoretic patterns that are influenced by analytical, biochemical, storage, and age factors. An understanding of the degree of influence exerted by each of these factors is necessary for the proper interpretation of test results. Dealing with those phenomena requires academic knowledge augmented by a good deal of practical experience.

The trend toward specialization is conspicuous and irreversible, and some practitioners are sincerely concerned that this development does not bode well for a profession that is consistently exposed to a diversity of specimens. Will the left hand know what the right hand is doing or is even capable of doing? Such concerns are understandable but surmountable. It is entirely appropriate if not compulsory that all personnel in the forensic laboratory be exposed to the workings of the entire facility. For example, one need not be a trained mass spectroscopist or a forensic serologist to comprehend the products of their labors. The fact is that skills used by all elements within a crime laboratory can and have been reduced to modest size textbooks, and it is not unreasonable to expect that any individual deemed qualified for employment in a forensic laboratory be capable of grasping and appreciating all its principles of operation. Prudent management dictates that laboratory personnel be rotated through each section of the laboratory and sensitized, if only for a brief time, to its workings. Laboratories large enough to compartmentalize their tasks must also put in place an interlocking system of supervision extending from the unit level to the laboratory director. This arrangement will formalize the lines of communication between laboratory specialists and assure a coordination of efforts.

In some ways forensic examiners have lost their mystique and sense of individual identity. The image of a lone examiner peering through a magnifying glass or a microscope has been replaced by a team of analysts working in laboratories filled with black (gray, tan, or green) boxes that are churning out graphs and digital displays. Actually, this modern image of forensic science is not totally accurate.

Technology Today and Tomorrow

Although it is true that over the past two decades, owing to developments in analytical instrumentation, there have been dramatic changes in the tools and techniques available to the forensic analyst, human involvement is still required to interpret and weigh the significance of data emanating from these machines. However, the point to emphasize is the discernible movement that has occurred toward the collection of data that are more objective than subjective in character. This trend is welcome and necessary if forensic science is to maintain credibility and sustain a reputation for impartiality. For example, it is disheartening, but understandable given the present state of the art, to watch eminent psychologists who have examined the same subject arrive at diametrically opposite views in the courtroom. Unfortunately, experience has shown that if an area of forensic science is totally dependent on the subjective skills of the analyst, then invariably "experts" will be found to express contradictory and equally persuasive views to a jury. If this situation occurs often enough and also happens to be the subject of contention in a number of highly publicized cases, the ultimate result will be a growing disillusionment with the credibility of science in the courtroom.

Fortunately, new technology has at least spared the physical and natural sciences from this fate. Although all areas of subjectivity have not yet been totally eliminated in the interpretation of forensic analytical data, the pace of progress is quickening. Even a mundane but subjective task such as comparing colors can now be performed with a microspectrophotometer. If necessary, the resultant spectral information can then be made available to experts, attorneys, and triers-of-fact alike to evaluate. Aided by like advances in analytical instrumentation, the search for more objective approaches for forensic analysis will continue.

Today, one can hardly pick up a newspaper or magazine without reading something about computers. The personal computer has literally brought the computer into the living room, but computers per se are not new to the analytical laboratory. Systems have been in existence for a number of years that are capable of collecting and storing voluminous quantities of laboratory data. However, the fact of the matter is that few forensic laboratories have invested in these costly computer systems. Obviously, the emergence of the low-cost computer is bound to change the complexity of the forensic science scene in ways that are difficult for us to envision. Ultimately, the likelihood is that minicomputers will link local crime laboratories via telephone lines to a large national computer. This realization will facilitate the collection of reference information on glass, paint, tireprints, shoeprints, and headlights. In the United States, a fragmented collection of local crime laboratories will be electronically merged into a single cohesive system. The computer will also have powerful consequences as far as the in-house interpretation of data is concerned. Chromatograms of accelerants, paints, and plastics, now manually compared and subjectively evaluated, will be scrutinized in precise fashion by pattern recognition algorithms. Comparisons between questioned and control materials will be judged and rated in exact and definable probabilistic terms, and not on subjective inclinations.

Perhaps these predictions are far removed from present reality, but they are not unreasonable or unrealistic. Advances in computerization and communications are being made at a far greater pace than anyone could have predicted just a few years ago. The forensic science community must begin to prepare for these eventualities. Standardization of techniques and procedures will become a necessary prerequisite if data and information are to be exchanged through computer networks. I hope that the technology that I envision will be as commonplace to forensic scientists of the twenty-first century as the scientific feats of Sherlock Holmes are to the forensic scientists of the 1980s.

5

Chemistry and the Challenge of Crime

Peter R. De Forest, Nicholas Petraco, and Lawrence Kobilinsky

Forensic science has been defined in many ways, but one of the most generally accepted definitions refers to forensic science as the application of the sciences to matters of the law. Forensic science is thus a broadly based field with many subspecialties. Forensic scientists approach their tasks by extracting knowledge and methods from the more established scientific disciplines, such as chemistry, physics, and biology, and adapting them to their own unique needs as well as developing new methods.

The discipline first began to develop in a formal way in the late 1800s. Although the role of the forensic scientist has changed little in the last one hundred years, he or she is no longer the "one man show" epitomized by the Sherlock Holmes character in Sir Arthur Conan Doyle's imaginative and prescient stories. Whereas a modified form of this "generalist" approach is valuable, with respect to complex modern problems, today's scientific investigator can only afford to master one of the various forensic disciplines, e.g., pathology, toxicology, odontology, questioned documents, and criminalistics.

Criminalistics is extremely broad in scope and is composed of many subdisciplines. The California Association of Criminalists has defined criminalistics as "that profession and scientific discipline directed to the recognition, identification, individualization, and evaluation of physical evidence by the application of the natural sciences to law–science matters." Criminalistics deals with the analysis of a wide variety of evidentiary materials. The criminalist relies most heavily on the principles and methods of chemistry. However, the most sophisticated analytical techniques are all too often inadequate for the more difficult tasks faced by the forensic scientist, and many challenging problems remain to be solved.

Problems confronted by various individual specialists in criminalistics can be very difficult and may require both general and in-depth knowledge in a number of areas, including microscopy, microchemistry, optical crys-

0784/83/0045$06.00/0
© American Chemical Society

tallography, instrumental methods of analysis, serology, immunology, genetics, physics, arson investigation, and crime scene reconstruction. Ideally, a criminalist should be a specialist in at least one or two areas and a generalist with respect to the overall field of criminalistics. Criminalists with complementary specialties often work together in an interdisciplinary team approach in order to analyze physical evidence and reconstruct a crime on the basis of the physical evidence. However, criminal law is not the only arena of forensic scientists; their scientific expertise and techniques are also applicable to civil investigations. The forensic scientist lends a vital service to the criminal justice system by providing meaningful scientific input for assisting the judicial system in determining the guilt or innocence of the accused in a criminal case.

Individualization and reconstruction are two of the most challenging tasks faced by the criminalist. *Individualization,* unique to criminalistics, involves using both physical and chemical information to evaluate the possibility that two items of evidence have a common origin. The individualization approach is called for when questions arise such as whether the paint from the scene of the hit-and-run originated from the suspect's car or whether the blood found on the suspect's clothing was that of the victim. *Reconstruction* involves analyzing all the information that is developed as a result of scientific investigation in the field and in the laboratory to determine details of the sequence of events that occurred at a crime scene. The role of the criminalist in this activity is similar to that of an archaeologist piecing together the mode of life and the habits of a past civilization based on an examination of the physical evidence unearthed at the site of a dig. Some of the activities of the criminalist and their intimate relation to chemistry will be the main focus of this chapter.

Historical Background

Historically, the first organized application of scientific disciplines to the field of criminal investigation dates back to Hans Gross, who in 1893 published his classical book, *Handbuch für Untersuchungsrichter.*[1] He introduced the word "Kriminalistic" from which the English word criminalistics is derived. Although he advocated a philosophy of drawing upon the expertise of various scientists to aid investigators and to supply proof for use in court, Gross was a magistrate rather than a scientist and did not contribute to the development of forensic science per se. By far the most significant contributions to the early development of forensic science methodology were made by Dr. Edmond Locard in France. In 1910 Dr. Locard became director of a scientific laboratory for the Lyons Police Department. During his years as director of the Institute of Criminalistics, Dr. Locard developed many new techniques. His contributions to the literature gained him worldwide recognition as one of the foremost criminalists of his day and further awakened police administrators and investigators to the value of and need for scientific evidence in any modern system of justice.[2]

In the early 1920s many forensic laboratories were established in various countries throughout Europe.[3] In the United States the first scientific laboratory devoted to the detection of crime was established in Los Angeles in 1923. Six years later a second crime laboratory at Northwestern University in Chicago became operational.[4] This laboratory was formed as a direct result of the crisis atmosphere generated by the St. Valentine's Day Massacre.[5] In the early 1930s crime detection laboratories were organized in New York City[6] and in Washington, D.C. by the Federal Bureau of Investigation.[7]

Criminalistics soon became a recognized academic discipline as a result of the scientific guidance and leadership of Dr. Paul L. Kirk at the University of California at Berkeley. Dr. Kirk's efforts resulted in a program devoted to training forensic scientists in the methods and technology of the established scientific community. He sought to encourage the application of the scientific method to solve the rather unique problems faced by the forensic community, and this has certainly not been an easy task.

Capabilities Today

Most of the applied scientific methods are primarily directed at the identification of a given substance such as a chemical compound, mineral, or plant. In contrast, forensic scientists are concerned with more than just the identification of a particular item of evidential material. Beyond identification, they must in many cases attempt to link the item to a unique or common source: they must try to *individualize* it. For example, suppose that the forensic scientist identifies a questioned fiber, which was removed from underneath the fingernails of a murder victim, as a blue nylon fiber. This identification alone is of little value unless this fiber can be associated with a unique source, e.g., a suspect's blue nylon shirt. The forensic scientist would like to be able to decide if the questioned blue fiber originated from the suspect's shirt or if it could have been derived from some other item or substance containing similar fibers. This task is extremely ambitious and difficult because the unequivocal linking of a particular item of evidence to a unique source is often not possible.

However, some police personnel, attorneys, and laymen, perhaps influenced by televised fictional presentations, believe that most items of evidential material can be individualized to a particular person, location, or thing. Many people believe that a fragment of glass, a paint chip, a length of rope, or a drug can be unequivocally associated with a unique or common source. Although it is undoubtedly true that all things are individual, we still lack the scientific sophistication for conclusive individualization in many cases.

At present, only a few types of physical evidence can be truly individualized. These include fingerprints, jigsaw matches, footwear and tire tread prints, striations present in tool marks, and ballistics evidence. Figures 1 and 2 illustrate two types of such evidence. Most other forms of evidence cannot be uniquely linked to a common origin as yet.

In the past, the available methods for the examination and comparison of the most often encountered types of evidential material (such as paint chips, glass fragments, hairs, fibers, drugs, physiological fluids, gunshot residue, explosive and arson materials) for the most part could only

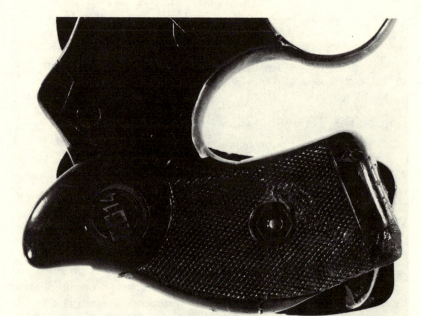

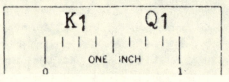

Figure 1. A jigsaw match of a broken gun hand grip.

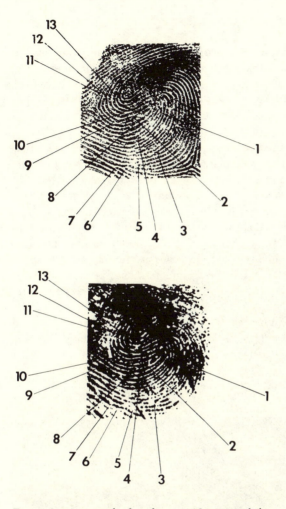

Figure 2. A court display showing 13 points of identity between the known inked print from the suspect (top) and the latent print (bottom) found at the crime scene.

enable the forensic scientist to determine whether the questioned and known evidential items were similar. If they were found significantly dissimilar, the questioned and known evidence did not share a common origin. However, in recent years, many new and improved instrumental and analytical methods allow the forensic scientist to come closer to the goal of associating items of physical evidence with a unique source.

Let us now consider some of these relatively new methods that have greatly extended the criminalists' capabilities.

Scanning Electron Microscopy

The scanning electron microscope (SEM) has found many applications in the field of crime detection. The instrument's capabilities of high magnification, high resolution, and extraordinary depth of field have made it the instrument of choice in many forensic examinations. In the past decade many published articles have advocated the use of the SEM in various forensic examinations.[8–15]

In the SEM, a focused beam of electrons is scanned over the surface of a specimen and causes secondary electrons to be emitted from the specimen. These secondary electrons are collected and detected to produce a signal that is displayed on a cathode ray tube (CRT). Magnification is simply the ratio of the size of the CRT scan to that of the sample scan and can be increased by reducing the size of the scanning pattern on the sample. Magnification ranges from $10\times$ to $200,000\times$ can be obtained.

The impact of the focused electron beam on the sample surface also causes x-rays to be emitted from the specimen. These x-rays are characteristic and can be used to identify the elements that are present in the specimen when the SEM is fitted with an x-ray microanalyzer. This technique of elemental analysis has proven especially useful in the area of gunshot residue (GSR) detection. When a firearm is discharged, a large cloud of residue is expelled (Figure 3). This residue contains substances

Figure 3. Cloud of gunshot residue that is evolved when a firearm is discharged. (Courtesy of the New York City Policy Laboratory.)

evolved from the primer, propellant charge, bullet, cartridge case, and lubricants. Furthermore, particles that are present have a morphology and elemental composition unique to GSR.[15] Currently, the SEM is used in the analysis to detect GSR by determining if these specific particles are present; however, a semiautomated SEM method needs to be developed so that GSR analyses can be performed on a routine basis.

Pyrolysis–Gas Chromatography

Pyrolysis–gas chromatography (PGC) has been shown to be a powerful analytical tool for the further characterization of synthetic fibers and paint samples in forensic science investigations.[16, 17] The potential value of PGC in forensic science has been recognized by forensic researchers over the past two decades.[16, 18–20]

Pyrolysis can be defined as the thermal degradation of intractable macromolecular materials into simpler units. The pyrolytic reaction is carried out in the absence of oxygen. In one type of PGC, a sample of fiber or paint (5–50 μg)* is placed in a quartz tube. The tube is then placed in contact with a coil of platinum wire. The sample is then inserted into the injection port of the gas chromatograph. The wire is electrically heated to the desired pyrolysis temperature in a few microseconds by feedback controlled circuitry. The volatile pyrolysis products that are produced are separated by gas chromatography (*see* box) into a characteristic *pyrogram*. For the purposes of identification, the characteristic pyrograms that are obtained have been likened to the fingerprint region of an infrared spectrum.[21] A standard method has been developed for the PGC analysis of synthetic fibers and paint polymers.[22]

Pyrolysis–gas chromatography, because of its high sensitivity and characteristic pyrograms, has proven to be a most valuable tool in the forensic characterization of synthetic fiber and paint specimens.

Gas Chromatography–Mass Spectrometry

In the war against narcotics, the combination of gas chromatography and mass spectrometry (GC–MS) is perhaps the most powerful weapon in the forensic scientist's arsenal. The ability of the gas chromatograph to separate complex mixtures into their components and the high sensitivity and specificity of the mass spectrometer have made this combination an indispensable tool.[23, 24]

The sample is prepared and injected into the gas chromatograph. The effluent is split by a variable splitter. Part of the effluent from the column goes to the flame ionization detector (FID) of the gas chroma-

---

*A microgram (μg) is 0.00000003 oz; therefore, 5–50 μg equals 0.00000015–0.0000015 oz, an exceedingly small amount!

> **Gas Chromatography**
>
> *Gas chromatography* is a technique for separating and analyzing compounds in complex mixtures. The sample to be analyzed, if not already a gas, is volatilized in the instrument by heat and picked up by a continuously flowing stream of inert gas. This carrier gas containing the sample is passed through a column containing a stationary phase to which the various sample components associate for different fractions of time as the carrier gas continues to flow through the column. The gas emerging from the column (the *effluent*) is monitored by a detection device coupled to a recorder that produces a chromatogram.

tograph. The remainder of the effluent goes to the separator interface, which separates the carrier gas and allows only the components of the mixture being examined to enter the ion source of the mass spectrometer. Here *ionization* (producing electrically charged molecules or atoms) of each component takes place, and the mass spectrum for each is recorded. Figure 4 shows a diagram of a GC–MS automatic data system. The data supplied by the GC–MS analysis allows the operator to make a rapid and unequivocal identification of a drug.

Even more important is the ability of the investigator to individualize the drug in question. Most street-level samples of narcotic drugs such as cocaine contain a large number of diluents. The actual quantity of narcotic drug present may be only a small fraction of the sample. The bulk of the material is usually a mixture of various commonly used cutting agents such as caffeine, sugar, quinine, lidocaine, and procaine. GC–MS has made it possible for the forensic scientist to analyze these complex mixtures quickly, and to give the narcotics officers information regarding their total composition. This information has been of great use to the intelligence sections of most drug enforcement agencies. These data have enabled forensic scientists to help narcotics investigators to trace illicit drug samples back to their sources and focus attention on the upper levels of the distribution system.

Thin Layer Chromatography

The development and worldwide distribution and use of textile fibers has made the forensic examination and comparison of fibers increasingly difficult. In a paper discussing the methods used in the forensic examination of hairs and fibers, Rash[25] suggested that further research be carried out on the development of methods for the extraction and comparison of dyes used in the process of dyeing synthetic fibers. A thin layer chromatography (TLC) method (*see* box) for this purpose has recently been developed.[26] The technique involves the extraction, separation, and com-

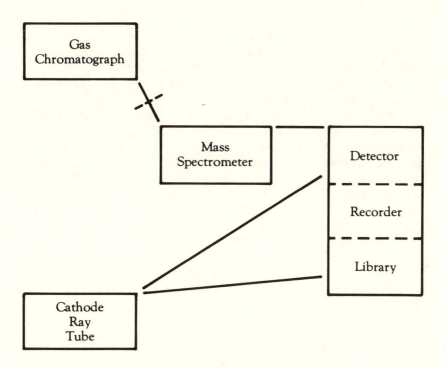

*Figure 4. Flow chart of a GC–MS automatic identification data
system.*

parison of dyes from generically similar fibers. This method has resulted
in increased discriminating power in the analysis of fibers that appear
similar by microscopic examination.[27]

In the TLC method, the dyes are extracted from a semimicro-sized
sample of textile fiber. This technique has the important advantage of

---

Thin Layer Chromatography

Thin layer chromatography is a technique for separating and analyzing
compounds in mixtures. Separation takes place in a thin layer of a porous
medium such as alumina or silica gel coated on a glass plate or an inert
plastic backing sheet. A small quantity of the sample mixture is placed at a
specific location near one end of the plate or sheet. This edge of the
assembly is placed in contact with a supply of a suitable "developing"
solvent, which wets the absorbent layer and begins to migrate across the
plate. The different constituents of the sample mixture are carried
through the layer at different rates and separate from one another as a
series of zones.

being nondestructive, which means that the fiber remains intact for further analysis or courtroom presentation. The extracted dyes are separated by using TLC, and the resulting dye patterns are compared with known dye patterns. Textile dyes are not pure compounds and will generally separate into one major colored component and one or more minor colored components. The successful separation and comparison of textile dyes by TLC will, therefore, depend to a large degree on the resolving power of the individual solvent systems employed. Figure 5 illustrates the separation of identical samples of acid orange 128 in five different solvent systems. Solvent system I separated the dye sample into two colored spots, whereas solvent system IV separated the identical dye sample into five colored spots. Because most dyed textile fibers contain a mixture of dyes, the superior resolving power of solvent system IV would result in an increase in discriminating power through the formation of a more complex multicolored dye pattern. This method has been employed by forensic scientists to discriminate among synthetic fibers that are the same generically and appear microscopically to be dyed the same color.

Fusion Microscopy

The study of the *morphology* (physical structure) and the *optical properties* (the way light is absorbed or refracted) of crystalline and pseudocrystalline samples as they are heated and cooled is known as fusion microscopy. The evidential value of textile fibers has been discussed by many forensic scientists.[28-31] Optical methods that can be used to characterize manmade fibers are essential to forensic examination. A recently developed fusion microscopic method[32] for synthetic fibers involves the observation of the changes in birefringence with temperature. *Birefringence* is the splitting of a light beam into two components that travel at different velocities.

A small length of fiber (5 mm) is mounted on a microscope slide in silicone oil. The specimen is placed on a hot stage. The fiber is examined with a polarized light microscope at the position of maximum brightness. The thickness is measured, and the birefringence is then determined. The temperature of the hot stage is raised 2 °C/min by the control unit. All changes are recorded at the temperatures at which they occur. The changes in birefringence are calculated from the data, and a plot of the changes in the logarithm of the birefringence versus temperature is made for the fiber specimen (Figure 6).

The graphs that are obtained have been useful in the identification and comparison of synthetic fibers, and in the differentiation among similar specimens of synthetic fibers that originated from different sources.[33]

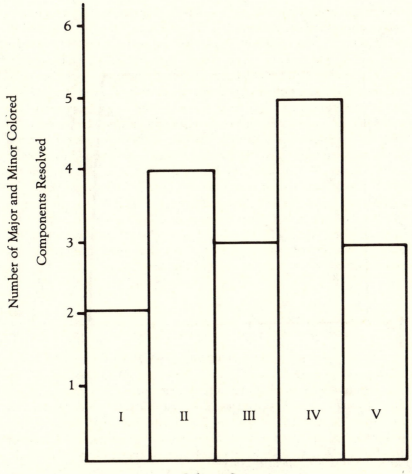

Figure 5. *Resolving power of five different solvent systems il-*
lustrated through separation of identical samples of
acid orange 128 (Du Pont, merpacyl orange R, lot
06666). (Courtesy of Ronald Resua, New York Po-
lice Laboratory.)

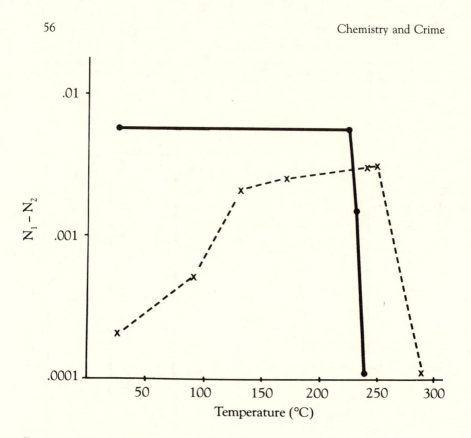

Figure 6. *Temperature versus birefringence curves for two types
of acetate fibers. Key: diacetate,* ──•; *and triacetate,*
── ─×.

Plasma Emission Spectroscopy

The plasma emission spectroscopy (PES) method uses the principles
of emission spectroscopy (*see* box) for the qualitative and quantitative
analysis of elements. Some PES instruments can be used in two basic
modes. The first mode, which is used for qualitative analysis, makes use
of an echelle grating and a polaroid camera attachment. The echelle
grating, using vertical as well as horizontal displacement, allows 28 wavelengths (1910–8000 Å) to be recorded on a 3 × 5-inch film. The location
of the elemental lines on this film is facilitated with the aid of a standard
template. This greatly simplifies the identification of the elemental composition (Figure 7). The second mode utilizes a spectrophotometric unit
that allows a specific element line to be focused and the intensity to be
measured, thus quantitating the element.

This instrument has been used in the elemental analysis of glass
samples for the purpose of categorizing the glass into various groups, such

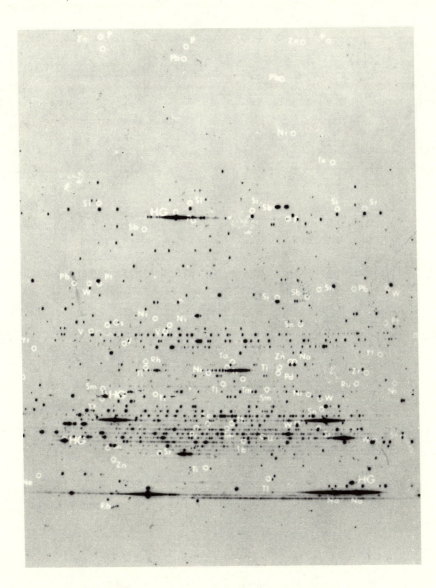

Figure 7. A photograph of the spectrum obtained by PES with the template used for element identification. (Courtesy of Tom Catalano, New York City Police Laboratory.)

> **Spectroscopy**
>
> *Spectroscopy* deals with the production, measurement, and interpretation of electromagnetic radiation resulting from emission or absorption of this radiation by matter (i.e., a sample of interest). *Emission spectra* are produced when radiant energy from the sample, excited by various forms of energy, is passed through a slit and then separated into a spectrum of various characteristic wavelengths by a prism or a grating.

as structural window, container, auto windshield, and headlamp (Table I). The rationale behind this approach is to support the existing physical property measurements test, such as density, refractive index, and dispersion by placing the glass into a smaller group for comparison. The ability to obtain qualitative information can also be very specific when unusual elements are found to be present within the questioned and known samples of glass. However, for the actual comparison of known and questioned glass samples, physical property measurements[34-45] still provide the highest degrees of discrimination in routine use in casework.

This method has been employed in actual casework. It has proven to be of great value in the categorizing of glass fragments into usage class, and in helping to discriminate among glass specimens with similar physical properties.

Serological Methods

Several years ago, the science of forensic serology entered a renaissance. As late as the mid 1960s the forensic serologist had reliable methods available only for accurate typing of the ABO blood grouping system in dried bloodstains.[46] The great majority of the population fell into the A or O groups, while the remainder constituted the B and AB groups (*see* Table II).

Because of the high frequency of the A and O types, the serologist's determination of ABO blood grouping often proved to be of little value in discriminating between blood specimens from different individuals. In fact, human red blood cells contain more than 160 types (groups) besides the ABO group, such as the MNSs and Rh groups. Blood types are also referred to as "genetic markers" or "antigens." With the development of procedures capable of discriminating more genetic markers, the serologist could produce a far more detailed description of an individual's blood. Although a bloodstain still cannot be individualized as a "fingerprint," it can often be discriminated to a frequency of occurrence in the population of greater than 1 in 50,000 or 0.002% of the population.[46]

Table I. Mean Percentage of Elemental Oxides Present in Various Types of Glass

Glass Usage	Al_2O_3	CaO	Fe_2O_3	MgO	NaO
Structural window	0.158	8.50	0.123	3.65	13.53
Auto windshield	0.150	8.10	0.555	3.93	12.90
Headlamp	1.370	0.017	0.060	0.017	5.40
Container	1.416	8.266	0.117	0.283	11.73

Courtesy of Tom Catalano, N.Y.C. Police Laboratory

Table II. Distribution of the Four Blood Types in the ABO Blood Grouping System in the U.S. Population

Blood Group	Percent of Population
A	40
O	42
B	15
AB	3

Radioimmunoassay

Although radioimmunoassay exists in many variations, the principle in most of these is that a competition is established between a fixed quantity of a radiolabeled antigenic substance* and an unknown amount of this same substance (usually present in some biological fluid or tissue) for binding to a fixed amount of monospecific antibody. If the biological sample lacks the substance, then all of the label will be bound to the antibody. If the sample contains an equal amount of substance to the labeled antigen, then 50% of the label will appear bound to the antibody; hence, the RIA is a quantitative assay with a sensitivity capable of detecting the antigen in the nanogram (1 billionth of a gram) or picogram (1 trillionth of a gram) range. In addition, RIA procedures are simple and specific, and their results are reproducible. This technique is important toxicologically as well as clinically. RIA procedures are currently available for assaying various drugs, protein and steroid hormones, and other biologically important substances.

In order to develop an RIA, a specific antiserum must first be produced. Drugs having a relatively low molecular weight are normally not antigenic; however, they can be coupled to a carrier protein and can be injected into rabbits as a *hapten–antigen conjugate* to raise a specific hy-

*See Chapter 9 for definitions of antigen, antibody, etc.

perimmune serum. Although this procedure requires the use of animals and usually requires a rather long period of immunization, it can be used to induce the production of antisera against virtually all compounds of clinical or forensic interest. Another consideration is that the rabbit will be a continuous source of the antisera for the remainder of its life.

The first report of the use of an RIA for drug detection was in 1970 by Spector and Parker.[47] They developed an RIA for morphine, which ultimately led to the development of RIAs for the detection of other opiates, barbiturates,[48] amphetamines,[49] LSD,[50] THC,[51,52] methadone,[53] and many other commonly abused drugs.[54]

An important advantage of RIA is that this procedure can be used when large numbers of samples must be assayed. An RIA has been developed and is currently being used to monitor the many patients on methadone maintenance programs.[55]

When small amounts of extremely potent drugs are ingested or injected, they become rapidly diluted to concentrations below the detection capabilities of most conventional assays. Thus, although thin layer chromatography and gas chromatography are quite sensitive and can be used for many drug assays, they are not as sensitive as RIA. In addition, in these procedures, the drug must first be extracted from the sample. GC can be time-consuming and is less useful when large numbers of samples must be examined in a relatively short time.

There has been a marked increase in the overuse and abuse of very dangerous drugs over the past decade. Having available a simple assay that can determine minute levels of a single drug or a combination of drugs in blood or urine is proving to be very helpful in both the clinical and criminal justice context.

In Summary

Thus we are left with "crime's challenge to chemistry." What this means is that we must continue to develop new methodologies and to increase the state of our scientific knowledge. This will one day enable the forensic scientist to approach more closely the goal of individualizing all forms of physical evidence left at a crime scene: an awesome challenge indeed, but one which must be met. For when this is accomplished, the forensic scientist will be able to fulfill his most important responsibilities: better reconstruction of crimes and helping to prove unequivocally the guilt or innocence of the accused.

References

[1] Walls, H. J. Forensic Science, An Introduction to Scientific Crime Detection, 2nd ed.; New York: Praeger, 1974; p. 1.

[2] Söderman, H.; O'Connell, J. J. *Modern Criminal Investigation*, 5th ed.; New York: Funk and Wagnalls, 1962; p. 58.

[3] Walls, H. J. *Forensic Science, An Introduction to Scientific Crime Detection*, 2nd ed.; New York: Praeger, 1974; p. 2.

[4] Goddard, C. *Am. J. Police Sci.* **1930**, 13.

[5] *Law, Medicine, Science and Justice* Bear, L., Ed.; Springfield, IL: Charles C. Thomas, 1964; p. 466.

[6] Söderman, H.; O'Connell, J. J. *Modern Criminal Investigation*, 5th ed.; New York: Funk and Wagnalls, 1962; p. 489.

[7] "Federal Bureau of Investigation, 40 Years of Distinguished Scientific Assistance to Law Enforcement," FBI Law Enforcement Bulletin **1972**, *41*, 3.

[8] Bradford, L. W.; Devaney, *J. J. Forensic Sci.* 1970, *15*, 110–19.

[9] Korda, E. J.; MacDonell, H. L.; Williams, J. P. *J. Crim. Law, Criminol. Police Sci.* **1970**, *61*, 453–58.

[10] Grove, C. A.; Judd, G.; Horn, R. *J. Forensic Sci.* **1972**, *17*, 645–58.

[11] Taylor, M. E. *J. Forensic Sci. Soc.* **1973**, *13*, 269–80.

[12] Garner, G. E. *J. Forensic Sci. Soc.* **1975**, *15*, 281–88.

[13] Solheim, T.; Leidal, T. I. *Forensic Sci.* **1975**, *6*, 205–15.

[14] Sabo, J.; Judd, G.; Ferriss, S. In "Forensic Science," Davies, Geoffrey, Ed.; ACS SYMPOSIUM SERIES No. 13, ACS: Washington, DC, 1975; pp. 75–82.

[15] Wolten, G. M.; Nesbitt, R. S.; Calloway, A. R.; Loper, G. L.; Jones, P. F. "Final Report on Particle Analysis for Gunshot Residue Detection," LEAA, U.S. Department of Justice, Washington, DC, 1977.

[16] Bortniak, J. P.; Brown, S. E.; Sild, E. H. *J. Forensic Sci.* **1971**, *16*, 380–92.

[17] Wheals, B. B.; Noble, W. *J. Forensic Sci. Soc.* **1974**, *14*, 23–31.

[18] Nelson, D. F.; Yee, J. L.; Kirk, P. L. *Microchem. J.* **1962**, *6*, 225–31.

[19] Jain, N. C.; Fontan, C. R.; Kirk, P. L. *J. Forensic Sci. Soc.* **1965**, *5*, 102–9.

[20] DeForest, P. R.; Kirk, P. L. *The Criminologist* **1973**, *8*, 35–45.

[21] The Textile Institute, *Identification of Textile Materials*, 7th ed.; 1975, p. 196.

[22] May, R. W.; Pearson, E. F.; Scothern, D. *Pyrolysis–Gas Chromatography* Analytical Sciences Monogram, No. 3, The Chemical Society, pp. 81–109.

[23] Klein, M.; Kruegel, A. V.; Sobol, S. P., Eds. "Instrumental Applications in Forensic Drug Chemistry," U.S. Department of Justice, DEA, Washington, DC, 1978, pp. 93–94.

[24] McNair, H. M.; Bonelli, E. J. *Basic Gas Chromatography*, 5th ed.; California: Varian, 1969; pp. 131–34.

[25] Rash, A. E. "Identification of Hairs and Fibers. A Review of the State of the Art," a paper prepared for the American Society of Crime Laboratory Directors, 1977, p. 5.

[26] Resua, R. *J. Forensic Sci.* **1980**, *25*, 168–73.

[27] Resua, R.; De Forest, P. R.; Harris, H. *J. Forensic Sci.* **1981**, *26*, 515–34.

[28] Plaa, G. L.; Barron, D. C.; Kirk, P. L. *J. Crim. Law, Criminol. Police Sci.* **1952**, *43*, 382–89.

[29] Kirk, P. L. *Crime Investigation: Physical Evidence and the Police Laboratory*; New York: Interscience, 1953; pp. 126–28.

[30] Longhetti, A.; Roche, G. W. *J. Forensic Sci.* **1958**, *3*, 303–29.

[31] Frei-Sulzer, M. *Methods Forensic Sci.*, Vol. IV; Curry, A. S., Ed.; New York: Interscience, 1965; pp. 141–76.

[32] Petraco, N.; De Forest, P. R.; Harris, H. *J. Forensic Sci.* **1980**,*25*, 571–82.

[33] *Ibid.*, pp. 575–81.

[34] Tryhorn, F. G. J. Crim. Law, Criminol. Police Sci. **1939**, 30, 404–19.

[35] Kirk, P. L. Density and Refractive Index; Springfield, IL: Charles C. Thomas, 1951; pp. 48–55.

[36] Ojena, S. M.; De Forest, P. R. J. Forensic Sci. Soc. **1972**, 12, 315–29.

[37] McCrone, W. C. J. Assoc. Off. Anal. Chem. **1973**, 56, 1223–26.

[38] McCrone, W. C. J. Assoc. Off. Anal. Chem. **1974**, 57, 668–70.

[39] McCrone, W. C.; Hudson, W. J. Forensic Sci. **1969**, 14, 370–82.

[40] Fong, W. J. Forensic Sci. Soc. **1971**, 11, 267–72.

[41] Nelson, D. F. In Methods Forensic Sci., Vol. IV; Curry, A. S., Ed.; New York: Interscience, 1965; pp. 138–39.

[42] Butterworth, A.; Gernan, B.; Morgans, D.; Scaplehorn, A. J. Forensic Sci. Soc. **1974**, 14, 41–44.

[43] Hughes, J. C.; Catterick, T.; Southeard, G. Forensic Sci. **1976**, 8, 217–27.

[44] Andrasko, J.; Maehly, A. C. J. Forensic Sci. **1978**, 23; 250–62.

[45] Catalano, T.; Sardone, J.; Harris, H. Crime Laboratory Digest **1981**, 1–3.

[46] Culliford, B. J. "The Examination and Typing of Bloodstains in the Crime Laboratory," U.S. Department of Justice, LEAA, Washington, DC, 1971; p. 16.

[47] Spector, S.; Parker, C. W. *Science* (*Washington, DC*) **1970**, *168*, 1347–48.

[48] Spector, S.; Flynn, E. J. *Science* (*Washington, DC*) **1971**, *174*, 1036–38.

[49] Cheng, L. T.; Kim, S. Y.; Chung, A.; Castro, A. *FEBS Lett.* **1973**, *36*, 339.

[50] Castro, A.; Grettie, D.; Bartos, F.; Bartos, D. *Res. Commun. Chem. Pathol. Pharmacol.* **1973**, *6*, 897.

[51] Teale, J. D.; Forman, E. J.; King, L. J.; Marks, V. *Proc. Soc. Anal. Chem.* **1974**, *11*, 219–20.

[52] Teale, J. D.; Forman, E. J.; King, L. J.; Piall, E. M.; Marks, V. *J. Pharm. Pharmacol.* **1975**, *27*, 465–72.

[53] Berman, A. R.; McGrath, J. P.; Permisohn, R. C.; Cella, J.A. *Clin. Chem.* (*Winston-Salem, N.C.*) **1975**, *21*, 1878.

[54] Cleeland, R.; Christenson, J.; Usategui-Gomez, M.; Heveran, J.; Davis, R.; Grunberg, E. *Clin. Chem.* (*Winston-Salem, N.C.*) **1976**, *22*, 712.

[55] Manning, T.; Bidanset, J. H.; Cohen, S.; Lukash, L. *J. Forensic Sci.* **1976**, *21*, 112.

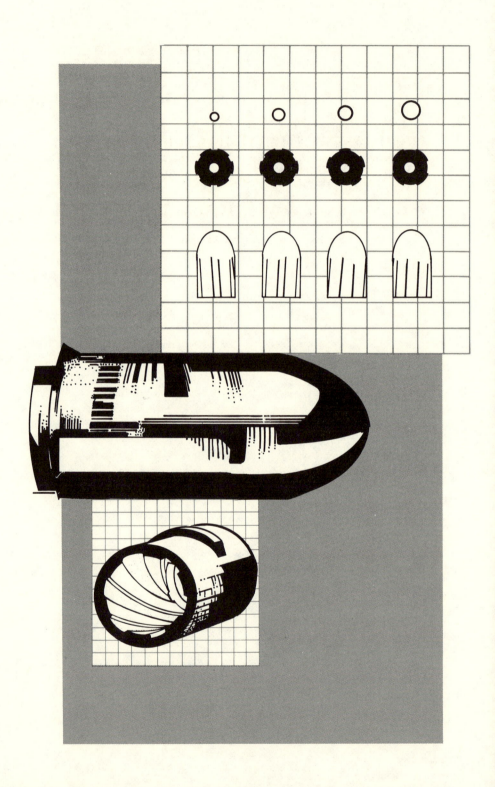

6

The Elemental Comparison of Bullet-Lead Evidence Specimens

Vincent P. Guinn

When a bullet is fired from a handgun (revolver or automatic pistol) or a rifle, it spins as it travels down the barrel, and the lands and grooves (Figure 1) of the striated gun barrel score the outer surface of the bullet. If the bullet is recovered in fairly good condition, a simple examination of it, with some magnification, will reveal the "class characteristics" of the gun barrel from which it was fired: number and dimensions of the lands and grooves, direction of twist (right or left), and angle of twist (in degrees). If such a bullet is then compared with a similar bullet test-fired in a particular gun (e.g., recovered from a suspect in a shooting case) in a comparison microscope, the detailed scratch marks ("tool marks") made on the bullet surfaces by the lands and grooves—if they line up exactly—can indicate to an experienced firearms examiner that both bullets were

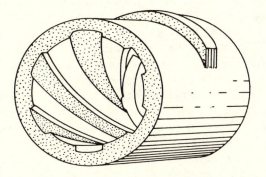

Figure 1. *Interior view of a gun barrel showing lands and grooves.*

0784/83/0065$06.00/0
© American Chemical Society

fired from the same gun barrel. If the two bullets are only the same in their class characteristics, but not in their tool marks, such a positive identification cannot be made.

In a great many shooting cases, the fatal or injury bullet is fragmented into various pieces or is severely deformed. Comparison microscope "matching" of the bullet to a particular recovered gun is then impossible, and often even the class characteristics cannot be determined. In such cases, an elemental analysis of the bullet fragments (or samples of a deformed bullet) and of the bullet portion of one or more unfired cartridges associated with a particular suspect can usually establish whether or not the bullet was produced from the same homogeneous batch of bullet lead as the bullet associated with the suspect.

The Manufacture of Bullet Leads

In general, commercial bullet leads may be put into two broad categories: soft lead and antimony-hardened lead. *Soft lead* may consist of highly purified virgin lead, highly purified or moderately purified scrap (reclaimed) lead, or mixtures of both. Soft lead may contain less than 1 ppm* (0.0001%) antimony (Sb) up to perhaps 1500 ppm (0.15%) antimony or even somewhat more. Harder leads are alloyed commercially with about 0.4% to 4% antimony—the more antimony, the harder the lead. Bullet/cartridge manufacturers usually specify maximum allowable concentrations for antimony and various undesirable impurity elements. Typically, they will accept hardened lead that is within about ±10% of the specified antimony concentration. In general, then, soft bullet leads are over 99.8% lead, and hardened bullet leads are roughly in the range of 95–99% lead.

Whether the bullet/cartridge manufacturer purchases bullet lead from a lead supplier or prepares his own bullet leads from lead and antimony, the procedure is generally roughly the following. A batch of lead of the desired composition (also variously referred to as a lot, a melt, or a heat of lead) is prepared by melting the desired amount of lead (usually in the range of 1–70 tons), raising the temperature to well above the melting point of lead (327 °C), adding the desired amount of antimony (in the case of hardened lead), allowing time for the antimony to dissolve in the molten lead and mixing well, then pouring the bullet lead out into 80–90 pound ingots (also variously referred to as billets or pigs of lead).

This description of the preparation of melts of bullet lead is simplified. Various producers, some or all of the time, employ different variations of the technique. In general, these variations result in smaller quantities of what is termed "homogeneous" melts of lead. Also, in the boxing of

---

*Ppm means parts per million parts by weight.

completed cartridges (discussed later) bullets from more than one homogeneous melt of lead may end up in the same box of cartridges, but this occurrence is not frequent. When it does occur, in the case of Sb-hardened bullets, all of the bullets in one box of cartridges will still have closely similar Sb concentrations (to meet specifications), but there may be two or even three measurably different combinations of Ag and Cu impurity concentrations.

To manufacture bullets of a specific size, shape, and composition, the bullet/cartridge manufacturer then processes a large number of these lead ingots in a device that extrudes the lead as bullet-lead "wire" of the desired diameter. This wire is then automatically cut into sections of the correct length to correspond with the desired bullet weight and squeezed into a mold of the desired bullet shape (either with or without a "copper jacket," as desired). Each formed bullet is then ejected into a storage bin. Later, the bullets are each fitted into a cartridge case that contains a primer and gunpowder, and crimped tightly in place. The complete cartridges are then boxed (usually in labeled boxes containing 20, 50, or 100 cartridges), and are ready for wholesale distribution.

Bullet weights are expressed in grains. One gram is equal to 15.432 grains or 0.035 ounce so a 10-gram bullet is a 154-grain or 0.35-ounce bullet. Bullets are made in the form of bare-lead bullets, semijacketed bullets, and "fully jacketed" bullets, and of various shapes (e.g., pointed, round-nose, hollow-point, "wadcutter") and sizes (weights and diameters). Semijacketed bullets have only the base of the bullet and part of the body of the bullet jacketed; some portion of the nose end is left unjacketed. "Fully jacketed" bullets have the entire lead core jacketed, except for the base end, where the bullet fits into the cartridge case. "Copper jackets" are actually made of brass; the most widely used compositions are 95% Cu/5% Zn, 90% Cu/10% Zn, or 87% Cu/13% Zn.

The main purpose of jacketing bullets is to reduce the "fouling" of gun barrels by the softer lead, with repeated use. If a fairly thick jacket is used, it also contributes to the mechanical strength of the bullet upon impact, making it possible to use soft lead in the cores of such bullets (although, commercially, even such heavily jacketed bullets often employ lead cores of antimony-hardened lead). Some military ammunition utilizes steel jackets (often nickel-plated), instead of copper jackets. Cartridge cases are usually made of a 70% Cu/30% Zn brass, in some cases nickel-plated to reduce corrosion.

The gunpowders used usually consist of particles of certain types and sizes, made up of a mixture of nitrocellulose and nitroglycerine. For all gun calibers except 0.22-caliber guns, each cartridge has a "center-fire primer" cup pressed into the center of the base of the cartridge (most 0.22-caliber cartridges instead employ a "rimfire primer"). When a cartridge is placed in the firing chamber of a gun and the trigger is squeezed, the firing pin impinges upon the shock-sensitive primer, causing it to detonate. When it detonates, it emits a jet of flame into the gunpowder, setting it off. Explosion of the gunpowder generates a very high pressure within the cartridge, causing the press-fitted bullet to separate from the cartridge case and accelerate down the barrel of the gun (spinning as it goes, due to the land-and-groove spirals) and out the muzzle of the gun.

With revolvers, the cartridges are loaded into a rotating cylinder. Typically, these cylinders hold up to six cartridges (hence the common term, "six-shooters"). After each cartridge is fired, the empty cartridge case remains in the cylinder, and the cylinder rotates so as to align the next cartridge with the base end of the gun barrel. With automatic pistols and rifles (except for single-shot varieties), the cartridges are loaded into a magazine, or clip. When one cartridge is fired, the empty cartridge case is ejected, and the next cartridge is moved into place in the firing chamber.

The Instrumental Neutron Activation Analysis of Bullet-Lead Specimens

The method that I use for analyzing bullet-lead specimens is called instrumental neutron activation analysis, or INAA. This method is also used by the FBI laboratory and various other law enforcement laboratories.

Beginning in 1962, my coworkers and I embarked on an extensive investigation of the possible applications of the method of neutron activation analysis (NAA) in the field of scientific crime investigation. These studies soon led to the development of the NAA method for the detection of primer gunshot residue on the back of the firing hand,[1,2] and to the INAA comparison of evidential specimens of bullet lead,[3,4] paint,[5] paper,[6] and other materials. Even small samples (usually 10–30 mg) of bullet lead can be analyzed rapidly, quantitatively, and nondestructively for their concentrations of Sb, Ag, Cu, As, and sometimes Sn.

Background samples of bullet leads of various known manufacturing origins (e.g., those marketed by Remington–Peters, Winchester–Western, Federal, Speer, and Sierra) and of various types and calibers, as well as evidential specimens from selected actual criminal cases, are analyzed in our laboratory by two INAA procedures. The first is a rapid-scanning procedure[7] (see box), and typically involves an irradiation time of 40 seconds, a decay time of 40 seconds, and a counting time of 40 seconds. Samples (each in a small polyethylene vial) are activated with neutrons in a nuclear reactor and then counted in a gamma-ray spectrometer in sequence, one at a time (along with standard samples of Sb, Ag, and Cu). The lower limits of this procedure are typically about 50 ppm Sb, 1 ppm Ag, and 10 ppm Cu. Because the great majority of bullet leads have concentrations of these elements that are a great deal higher than these

Neutron Activation Analysis

In *neutron activation analysis*, a specimen is irradiated with neutrons; this irradiation causes the specimen to become radioactive; that is, it emits gamma-rays. These rays can be measured in an instrument called a spectrometer. A *neutron* is an elementary particle that has no electric charge; it is a constituent of all atoms.

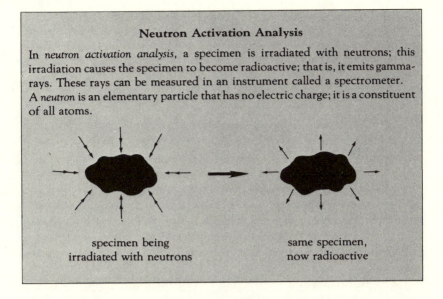

specimen being
irradiated with neutrons

same specimen,
now radioactive

levels, the rapid-screening method is usually all that is needed to determine these three elements to quite good precisions.

However, in those instances in which (1) the rapid-screening method does not provide adequate measurement precisions for Sb or Cu (due to unusually low concentrations, or instances in which less than 1 mg of sample is available), or (2) measurement of the As concentration is desired, a longer procedure is used.[8] In our laboratory, this procedure[8] typically involves the simultaneous irradiation in the reactor of up to 40 samples and standards for 1 hour. Then, commencing at decay times of the order of one to a few hours, each activated sample is counted on the gamma-ray spectrometer, in succession, for 5 or 10 minutes. This longer procedure detects and measures three elements in bullet-lead samples (Sb, Cu, and As), but not Ag.

A fifth element, tin, can sometimes be detected and measured in bullet-lead samples by INAA, but with less accuracy than the other four elements, and it requires an irradiation time/decay time combination in between that used in the rapid-screening method and the longer method.

In general, if the rapid-screening procedure reveals marked differences in the elemental composition (Sb, Ag, and Cu concentrations) of two bullet-lead samples (e.g., a sample from a fatal bullet, and one from a cartridge found in the possession of a suspect), it is apparent that they were not produced from the same homogeneous melt of lead, and hence no further analysis is necessary. If, however, the two samples being compared are analytically indistinguishable from one another in their Sb, Ag, and Cu concentrations, it is desirable to also compare them via their As concentrations by using the longer procedure to provide four points of comparison instead of just three. In crucial cases, it is even worthwhile to use a third (intermediate) INAA procedure, in an effort to detect and measure a fifth element, Sn. If successful, this measurement would then provide a total of five points of comparison for the decision as to whether the two samples came from the same homogeneous melt of lead.

Because of an inadequate background file of bullet-lead compositions and for other reasons, it is not possible to make an accurate calculation of the mathematical probability that two analytically indistinguishable bullet-lead samples were produced from the same homogeneous melt of lead. Instead, one must resort to more qualitative expressions, such as "probably" (if only three elements were measured), "very probably" (if four elements were measured), or "highly probably" (if five elements were measured). Depending in individual cases on how relatively common or uncommon the observed concentrations are (among the whole population of bullet leads) and on how accurately and precisely each concentration was measured, these three qualitative expressions of probability of a common melt origin may correspond, respectively, to probabilities of the order of perhaps 99, 99.9, and 99.99%.

It is also feasible, if necessary, to detect and measure elements besides Sb, Ag, Cu, As, and Sn in bullet-lead samples if one is willing to consume perhaps 10 mg of sample and to spend the additional time necessary.

In the FBI laboratory, the longer INAA procedure is employed, but not the rapid-screening procedure. As a result, they routinely determine Sb, Cu, and As in bullet-lead samples. Unfortunately, many samples are too low in As for very precise measurement. Another difficulty is that bullet fragments and samples taken from mashed bullets often have bits of copper jacket imbedded or buried in them (if the bullet was a copper-jacketed bullet), thus resulting in spuriously high measured copper concentrations. Of course, such jacket contamination of the sample also produces erroneously high copper values in the rapid-screening INAA procedure. Whenever such useless copper values are encountered, the longer INAA procedure reduces to just two useful elements (Sb and As, if the As concentration is high enough), and the rapid-screening INAA procedure also reduces to just two useful elements (Sb and Ag). In such cases of copper contamination, it is especially desirable to use both INAA procedures to determine a possible total of three useful comparison elements (Sb, Ag, and As).

In preparing bullet-lead samples for INAA measurement, specimens of bullet fragments or fired bullets from the crime scene or victims (usually designated as "Q" samples), and pieces cut from bullets of unfired cartridges connected with a suspect (usually designated as "K" samples, since their brand is known), are examined under magnification to ascertain whether there is any visible evidence of adhering jacket material. If there is, one attempts to remove the jacket material with a surgical scalpel. In our laboratory, such samples are then further processed by immersing each sample in concentrated nitric acid for 10 minutes at room temperature.[9] This procedure will dissolve away any specks of adhering jacket material without dissolving any measurable amount of the lead material. However, even this acid-treatment procedure fails if there are jacket particles completely imbedded in the lead and inaccessible to attack by the nitric acid.

Bullet-lead elemental compositions presented in court in criminal cases usually include only three elements: Sb, Cu, and As (if analyzed by the FBI laboratory) or Sb, Ag, and Cu (if analyzed in our laboratory by the INAA rapid-screening procedure).

Applications in Some Illustrative Criminal Cases

INAA has been used, in toto, for the analysis of several thousand bullet-lead evidence specimens (of both unknown and known brand origins), involved in many hundreds of criminal cases. The results of such analyses and their interpretations have been presented in U.S. courts in hundreds of cases. A few illustrative, rather well known, cases in which I conducted the analyses are briefly summarized as follows.

The SLA Shootout

On May 17, 1974, six members of the "Symbionese Liberation Army" barricaded themselves in a house in Los Angeles, and engaged in a wild shootout with Los Angeles Police Department SWAT members. In the course of the shootout, each side fired some 4,000 to 5,000 rounds of ammunition. Finally, the house caught fire, but firing from the house continued for a time. After the fire had been put out, the badly burned bodies of the six SLA members were found in the ruins. Autopsies performed in the Los Angeles County Coroner's Office established the identities of the six bodies, and their causes of death (gunshot wounds or fire). The leader of the SLA, so-called "Marshall Cinque" (his real name was Donald De Freeze) was found to have died from a through-and-through bullet wound to the head—in one side, out the other. The fatal bullet was never recovered, but along the wound track in the head the bullet left quite a number of small pieces of bullet lead because it struck bone.

For a variety of reasons, it was felt necessary to establish whether De Freeze had been killed by a SWAT bullet or had committed suicide as the gun battle drew to a close and the fire in the house spread. We first analyzed ten of the small bullet-lead fragments found along the wound track in De Freeze's brain, replicate samples of bullet lead from the three kinds of ammunition he had been firing (0.38-caliber revolver ammunition, unexploded cartridges found by his body), and replicate samples of bullet lead from some twelve kinds and brands of ammunition used by SWAT in the gun fight. The rapid-screening INAA method quickly eliminated all twelve kinds of SWAT ammunition, and two of the three kinds of ammunition that De Freeze had been firing. At this point, it appeared likely that he had committed suicide.

However, concurrent measurements of wound-track particles, conducted by Ronald L. Taylor at the Los Angeles County Coroner's Laboratory, revealed that the fatal bullet was a nickel-plated steel-jacketed bullet. Because all of the SWAT ammunition and all of De Freeze's ammunition was copper-jacketed, this finding was indeed puzzling. Then, information was received that one of the SWAT members had been firing nonregulation World War II surplus 9-mm military ammunition, steel-jacketed. A search revealed that thirteen different kinds of World War II surplus 9-mm military ammunition were widely available in the United States. All of these were jacketed with nickel-plated steel.

Specimens of all thirteen of these foreign-made cartridges were obtained, and we analyzed their bullet leads by INAA. Twelve of them showed markedly different elemental compositions than that of the fatal bullet fragments, but one showed essentially an identical composition. This one was from a cartridge manufactured in Czechoslovakia in 1941 (during the German occupation). Both the fatal bullet fragments and this one were of very high purity, soft lead, containing only a few ppm each of Sb, Ag, and Cu.

The final conclusion was that De Freeze did not commit suicide, but instead was killed by a nonregulation 9-mm steel-jacketed SWAT bullet, probably manufactured in Czechoslovakia in 1941.

The Oscar Bonavena Homicide

On May 22, 1976, Argentine heavyweight boxer Oscar Bonavena was shot and killed at the Mustang Ranch in Nevada, the biggest and most "celebrated" legal brothel in the United States. He was hit by a single bullet, which passed through his body and was never recovered. However, at autopsy (performed for the Nevada authorities by the Los Angeles County Coroner's Office), many small fragments of the fatal bullet and one of bullet-jacket material were found along the wound track. Measurements by Ronald L. Taylor revealed that the bullet-lead fragments were of antimony-hardened lead, and that the jacket material was a brass of approximately 95% Cu/5% Zn composition.

Witnesses observed two men firing at Oscar Bonavena with rifles: one with a 30.06 rifle, the other with an AR-15 rifle (0.223-caliber ammunition). Both men fled the scene, discarding the rifles. Four unfired cartridges were found in the 30.06 rifle, and two in the AR-15 rifle. Measurements by Ronald L. Taylor revealed that the 30.06 bullets were of antimony-hardened lead and they were jacketed in 95% Cu/5% Zn brass. Taylor could not detect antimony in the 0.223-caliber bullet lead, but could show that they were jacketed in 90% Cu/10% Zn brass.

We then analyzed ten of the fragments of the fatal bullet, the four recovered 30.06 bullets, and the two 0.223-caliber bullets by INAA. The results were quite conclusive, and further confirmed Taylor's results. The ten fragments of the fatal bullet averaged 4.85 ± 0.29% Sb, 44 ± 6 ppm Ag, and 780 ± 120 ppm Cu, and the four 30.06 bullets averaged 4.72 ± 0.15% Sb, 38 ± 4 ppm Ag, and 470 ± 300 ppm Cu. By contrast, the two 0.223-caliber bullets averaged 0.75 ± 0.03% Sb, 71 ± 2 ppm Ag, and 300 ± 40 ppm Cu. Because one of these two kinds of ammunition definitely killed Oscar Bonavena, it was clearly the 30.06 ammunition that did so. The man who was firing at Bonavena with the 30.06 rifle was identified as Ross Brymer, a bodyguard of the brothel's owner. Despite the solid first-degree murder evidence against him, he was not prosecuted for first degree murder. Instead, much later, he pleaded guilty to a lesser charge, was given a two-year sentence for voluntary manslaughter, and served an even shorter term in prison.

The Assassination of President John F. Kennedy

On November 22, 1963, President Kennedy was shot and killed by rifle fire in Dallas during a motorcade. Shortly thereafter, the suspected assassin Lee Harvey Oswald was captured, but only after he had shot and killed a Dallas police officer who was attempting to arrest him. Oswald was seen to shoot the officer with a 0.38-caliber revolver. On the sixth floor of the Texas Book Depository building, from whence the rifle shots had come, police found a Mannlicher-Carcano (MC) 6.5-mm rifle (with one unfired cartridge still in it) and three spent Western Cartridge Company (WCC) 6.5-mm MC cartridge cases. The consensus of opinion among witnesses was that three rifle shots emanated from that room when the President and Texas Governor John Connally were hit. Oswald was never brought to trial for the assassination, because, two days later, while being transferred by police, he himself was shot and killed by Jack Ruby.

The new President, Lyndon B. Johnson, appointed a commission, headed by U.S. Supreme Court Chief Justice Earl Warren, to conduct an in-depth investigation of the assassination. In the fall of 1964, the Warren Commission issued its report[10] on its findings. They concluded that President Kennedy had been killed and Governor Connally seriously injured by MC bullets fired from the Book Depository building by Oswald. They also concluded that there was no solid evidence that anyone else fired any shots at the time.

The Warren Commission Report generated a great deal of discussion, speculation, and disagreement, and resulted in numerous books, mostly disagreeing with at least various parts of the Commission Report and proposing other theories as to who fired the shots that killed the President.

Some years later, we reviewed all of the FBI's data that pertained to the MC rifle, the MC ammunition, and their INAA and other examinations of the various recovered bullet-lead fragments and portions of the bullet jackets. The results (Sb and Ag only) showed that the various bullet-lead fragments recovered from the victims and the limousine were generally similar in composition to the WCC 6.5-mm MC type of bullet lead, but could not establish for sure whether or not some additional kind of ammunition might have struck, or whether or not more than two bullets struck. Microscopic examination did establish quite firmly that the "stretcher bullet" (an only slightly damaged bullet found on Governor Connally's stretcher, and termed by some the "magic bullet") had been fired from Oswald's rifle, and that the recovered nose-end portion of a bullet jacket and the recovered base-end portion of a bullet jacket (recovered in the limousine) were both fired from his rifle.

In 1977, the U.S. House of Representatives appointed a Select Committee on Assassinations to conduct intensive reinvestigations of the assassinations of President Kennedy and of Martin Luther King, Jr. The Select Committee asked me if I would agree to reanalyze all of the bullet-lead evidence specimens involved in the Kennedy assassination, again by INAA but using more modern equipment.

Details of my new, and this time definitive, analyses of the bullet-lead specimens have been reported elsewhere,[11] so only the results will be repeated here. They showed that specimen CE–399* (the Connally stretcher bullet) and specimen CE–842 (fragments recovered from Connally's shattered right wrist) closely matched one another in their Sb and Ag concentrations: mean values of 815 ± 25 ppm Sb and 9.30 ± 0.71 ppm Ag. The other three specimens analyzed, CE–567 (a large fragment found in the car), CE–843 (fragments recovered from Kennedy's brain), and CE–840 (small fragments found in the car), agreed closely with one another in their Sb and Ag concentrations: mean values of 622 ± 20 ppm Sb and 8.07 ± 0.15 ppm Ag. These mean values are markedly lower in Sb (622 versus 815 ppm) and somewhat lower in Ag (8.07 versus 9.30 ppm) than the mean values for specimens CE–399 and CE–842. Clearly, the results showed the presence of two bullet leads of analytically quite distinguishable compositions, and showed no evidence for the presence of more than two bullet leads. Copper was also determined in all five specimens, but the results were not so clear: the Connally stretcher bullet (CE–399) showed a Cu value of 58 ± 3 ppm, whereas the CE–567, CE–843, and CE–840 specimens showed a mean value of 41 ± 1.7 ppm Cu. The fragments from Governor Connally's wrist (CE–842) showed a very high Cu concentration

---

*The CE numbers are the original Warren Commission exhibit identification numbers.

(994 ppm), indicating that they were probably contaminated with imbedded copper jacket material, and hence invalidating the usefulness of the copper value. Incidentally, the INAA results also showed that all of the specimens were largely composed of lead, the mean Pb concentration (not measurable with high precision) being $98.7 \pm 3.9\%$.

In this case, it was possible to distinguish by INAA the presence of two bullet leads, even though of the same brand and fired on one occasion from the same rifle, and hence probably taken from the same box (of 20) WCC cartridges by Oswald. Normally, for most brands of ammunition, all of the bullets of the cartridges in a given box of cartridges are analytically indistinguishable from one another. However, the history of these WCC 6.5-mm MC cartridges is strange. Four million of them were produced in 1954 by WCC for the U.S. Army. Apparently, they were intended to be used outside of the United States, by others, because the U.S. Army does not use 6.5-mm Mannlicher-Carcano rifles (an Italian World War II military rifle). At some time prior to the 1963 assassination, sizable quantities of this ammunition reentered the United States and were sold in war surplus stores. At least much of it arrived back still in the original WCC cardboard boxes of 20, but packed into crates coming from Greece.

At some point in their history, the bullets produced in the many melts of lead used by WCC to manufacture the four million cartridges were pretty thoroughly mixed in the boxes. I had established this fact[12] some years before the 1977 reinvestigation study. In measurements of many samples taken from various purchased boxes of this ammunition, many different bullet-lead compositions were found in any one box of cartridges. Although all met the original U.S. Army specification of soft lead ($\geqslant 99.85\%$ lead), their Sb concentrations ranged from 15 to 1200 ppm, the Ag concentrations from 5 to 22 ppm, and the Cu concentrations from 10 to 370 ppm.

I also disassembled the one unfired MC 6.5-mm WCC cartridge (CE–141) found in Oswald's rifle, took a small sample of its bullet lead, then reassembled the cartridge. This bullet sample was then analyzed by INAA and found to be quite different in elemental composition from the other two MC bullets fired by Oswald, being only 15 ppm Sb, but 22 ppm Ag, and 22 ppm Cu. This bullet had not been analyzed previously. The WCC 6.5-mm bullets were heavily jacketed (3.30 grams of 90% Cu/10% Zn, completely surrounding a 7.13-gram lead core, except at the open base of the bullet).

In September of 1978, I presented my findings at the Public Hearings held in Washington, D.C. by the Select Committee. These findings, which might have disagreed with the original Warren Commission hypothesis of only one man (Oswald), and two Oswald MC bullets that struck occupants of the limousine, instead served to reinforce their original conclusion. The other bullet fired by Oswald (very likely the first one he fired) apparently missed its mark and was never found—just its empty cartridge case in the book depository building. His second shot struck the President in the back, exited from his throat, entered the Governor's back, exited from his chest, shattered his right wrist, and then, nearly spent, slightly penetrated his left thigh, to fall out on his stretcher at the hospital. This bullet left no particles along the wound track in either the President or the Governor, and hence was not damaged (even though it broke one of the Governor's ribs in a glancing blow) until it struck the Governor's right wrist. Here, it suffered a dent in its nose and it lost about 1% of its lead [recovered as several small pieces (CE–842)]. Oswald's third and final shot, the fatal shot, struck the President in the back of the head and exited in the right front of the head, the bullet fracturing into a number of pieces (CE–567, CE–840, and CE–843).

My findings, of course, neither prove nor disprove the various conspiracy speculations, such as someone, in addition to Oswald, firing from some other location such as the "grassy knoll." They do show that if any other persons were firing, they did not hit anyone or anything in the President's limousine.

References

[1] Ruch, R. R.; Guinn, V. P.; Pinker, R. H. Trans. Am. Nucl. Soc. 1962, 5, 282 (followed by more extensive paper in Nucl. Sci. Eng. 1964, 20, 381–85).

[2] Schlesinger, H. L.; Lukens, H. R.; Guinn, V. P.; Hackleman, R. P.; Korts, R. F. "Special Report on Gunshot Residues Measured by Neutron Activation Analysis," U.S. Atomic Energy Commission Report GA-9829, 1970, 144 pp.

[3] Lukens, H. R.; Guinn, V. P. J. Forensic Sci. 1971, 16, 301–8.

[4] Lukens, H. R.; Schlesinger, H. L.; Guinn, V. P.; Hackleman, R. P. U.S. Atomic Energy Commission Report GA–10141, 1970, 48 pp.

[5] Schlesinger, H. L.; Lukens, H. R.; Bryan, D. E.; Guinn, V. P.; Hackleman, R. P. U.S. Atomic Energy Commission Report GA–10142, 1970, 261 pp.

[6] Lukens, H. R.; Schlesinger, H. L.; Settle, D. M.; Guinn, V. P. U.S. Atomic Energy Commission Report GA–10113, 1970, 50 pp.

[7] Guinn, V. P.; Purcell, M. A. J. Radioanal. Chem. 1977, 39, 85–91.

[8] Guinn, V. P. J. Radioanal. Chem. 1982, 72, 645–64.

[9] Izak-Biran, T.; Guinn, V. P.; Purcell, M. A. J. Forensic Sci. 1980, 25, 374–79.

[10] Warren Commission, "Report of the President's Commission on the Assassination of President John F. Kennedy," U.S. Government Printing Office: Washington, DC, 1964, 888 pp.

[11] Guinn, V. P. *Anal. Chem.* **1979**, *51*, 484A–493A.

[12] Izak-Biran, T.; Guinn, V.P. *Trans. Am. Nucl. Soc.* **1978**, *28*, 94.

[13] Guinn, V. P.; Izak-Biran, T.; Purcell, M. A.; Cassorla, V.; Nichols, J. *Trans. Am. Nucl. Soc.* **1979**, *32*, 188–89. (*See also* Guinn, V. P.; Nichols, J. *Trans. Am. Nucl. Soc.* **1978**, *28*, 92–93.)

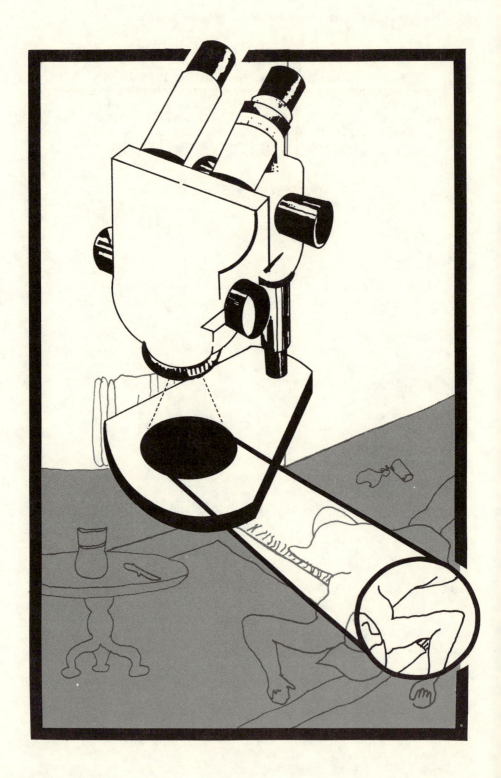

7

Bloodstain Analysis

Case Histories

Frances M. Gdowski

It is 1965. A man is apprehended by the police two miles from the scene of a homicide. There are bloodstains on the man's shirtsleeve; the police suspect that the blood came from the victim. The man tells the police that he was in a fight two weeks before and it is his own blood on his clothing. Later, the bloodstain is analyzed and found to be Group O; both the victim and suspect are Group O. The fact that all three blood groupings are the same prevents the police from either including or excluding the apprehended man as the suspect.

Background

Until the early 1970s, analysis of bloodstain evidence in the United States was limited to three basic types of testing: (1) identifying a stain as blood, (2) identifying the stain as being of human or animal origin, and (3) identifying the blood group. These analyses were of limited use to the investigator because blood groups such as A and O occur in a sizable portion of the population (Table I), and if both the victim and the suspect have the same blood group, one sample could not be distinguished from the other.

Table I. United States Blood Group Frequencies (percent)

Blood Group	White	Black
O	45	49
A	40	27
B	11	20
AB	4	4

SOURCE: Ref. 1.

0784/83/0079$06.00/0
© American Chemical Society

In 1967, Brian Culliford of the Metropolitan Police Laboratory found that the enzyme phosphoglucomutase (PGM_1) could be detected in dried bloodstains.[2] This finding proved useful to the bloodstain analyst as well as the forensic investigator because this enzyme is *polymorphic** and it occurs in the population in frequencies that could be useful to the investigator (Table II).

Since then, several *genetic markers*, including both enzymes and serum proteins, have been found to be detectable in dried bloodstains and have been used in forensic analysis (Table III). In addition, several of the genetic markers have been found to be polymorphic in a specific racial group; one example is transferrin, a serum protein in which the phenotype CD is rare in whites, but occurs in 8–10% of the black population.[4]

How does this type of analysis aid the investigator?

---

*See Glossary on p. 85.

Table II. New Jersey Phosphoglucomutase (PGM_1) Frequencies (percent)

Phenotype	White	Black
1	59.9	66.5
2-1	35.0	28.0
2	5.1	5.5

SOURCE: Ref. 3.

Table III. Representative Sample of Genetic Markers Identified in Bloodstains

Genetic Marker	Number of Phenotypes
Adenosine deaminase (ADA)	3
Adenylate kinase (AK)	3
Carbonic anhydrase II (CA II)	3
Erythrocyte acid phosphatase (ACP)	6
Esterase D (EsD)	3
Glyoxalase I (GLO I)	3
Group-specific component (Gc)	3
Haptoglobin (Hp)	3
Phosphoglucomutase (PGM_1)	3
Transferrin (Tf)	2

- It provides a genetic profile of both the victim and the suspect.

- It increases the discrimination potential of a specific bloodstain because it lowers the frequency of occurrence of the stain in a given population.

- It provides the capability to distinguish between two persons of the same blood group.

- It can, in some instances, indicate the possible racial origin of the bloodstain.

The value of these genetic analyses is illustrated in the following three case histories.

Case I—E.S. Rape

In November of 1979, a woman (E.S.) reported to the police that she had been raped. She stated that her assailant was black and had gained entrance into her third-story bedroom by breaking through the glass door leading from her balcony. The man had pulled the phone from the wall, threatened the woman, then raped her. Blood from a cut on the man's hand was left on the victim's nightgown. The nightgown and a sample of the victim's blood were submitted to the laboratory for analysis.

The victim's blood was analyzed and was shown to be Group O and to exhibit the genetic markers shown in the top row in Table IV. Next, the bloodstain on the nightgown was analyzed and the results were compared to the victim's blood control (middle row, Table IV). Although both the victim's blood and the bloodstain were Group O, the bloodstain did not originate from the victim because the other genetic marker phenotypes were different. The identification of the Tf phenotype CD in the stain from the nightgown indicated that the stain could have come from a black person, thus corroborating the victim's statement that she was raped by a black man.

Calculating the frequency of occurrence in the black population of blood Group O and the genetic markers identified in the bloodstain indicates that 0.3% of the population would be expected to have these phenotypes; or, 99.7% of the black population was eliminated as being the source of this bloodstain.

Case II—E.S. Homicide

Three weeks after she had been raped, E.S. was found brutally murdered in her bedroom. The bedroom was strewn with bloodstained items and burnt newspapers; the phone had been ripped from the wall.

Table IV. E.S. Rape—Blood Analysis Results

Specimen	Blood Group	Genetic Marker Phenotype*									
		EsD	PGM	GLO I	CA II	ADA	ACP	AK	Hp	Gc	Tf
E.S. blood control	O	2-1	2-1	2-1	1	1	BA	2-1	2	2-1	C
Stain from nightgown	O	1	1	—	1	1	B	1	2-1	1	CD
K.A. blood control	O	1	1	2-1	1	1	B	1	2-1	1	CD

*See Table III for the names of the genetic markers.

The perpetrator appeared to have entered her bedroom through the broken pane in the balcony door. A search of the balcony was conducted and a bloodstained red bandana was collected. This item and other bloodstained specimens were submitted to the laboratory for analysis. Comparisons were made of the blood from the scene with the victim's blood to determine the locations and instruments of the assault.

Case III—A.C. Homicide

Approximately two weeks after the E.S. homicide, a second woman was found brutally murdered in the same city. Her body was found in the upstairs hallway in a condition similar to that of E.S. Her bankbook was found in the toilet; her car and credit cards were stolen.

Two days later an off-duty patrolman identified the woman's car parked outside a sporting goods store. He waited until a man emerged from the store and unlocked the car. The officer then arrested him. The defendant's clothing was bloodstained, he had the victim's credit cards in his possession, and he was black.

The victim's and suspect's blood controls were analyzed; the genetic markers present in each control were compared with the genetic markers identified from the suspect's bloodstained clothing. The genetic markers identified indicated that the blood on the suspect's clothing did not originate from the suspect and could have originated from the victim (Table V).

Because of the brutality of both the A.C. and E.S. homicides, the police suspected that the defendant, K.A., was involved in the E.S. murder. When the suspect was questioned, he strongly denied having any knowledge of the E.S. murder or rape, but the analysis of the physical evidence indicated otherwise (Table IV). When the blood group and genetic markers of the bloodstain from the nightgown in the E.S. rape were compared with the blood group and genetic markers identified in

Table V. A.C. Homicide—Blood Analysis Results

Specimen	Blood Group	EsD	PGM	GLO I	CA II	ADA	ACP	AK	Hp	Gc	Tf
A.C. blood control (V)	A	—	1	2	1	1	BA	1	1	1	C
K.A. blood controls (S)	O	1	1	2-1	1	1	B	1	2-1	1	CD
Suspect's clothing											
Undershirt, left front	—	1	1	2	1	1	—	1	—	—	—
Undershirt, left cuff	A	1	1	2	—	1	BA	1	—	—	—
Dungarees, right cuff	—	1	1	—	1	1	BA	1	1	—	C
Jacket, right cuff	—	1	—	—	—	1	BA	1	—	—	—
Sweatshirt, left cuff	—	1	—	—	—	1	—	1	1	1	C

the suspect's blood control, both the blood group and the markers were found to be the same, indicating that K.A. may have been the perpetrator. K.A. was again questioned regarding the E.S. homicide and denied being at the scene. The suspect's blood was then compared to physical evidence collected from the E.S. homicide (Table VI).

The specimens collected from the victim's bedroom excluded the suspect as the source of the blood; however, the genetic markers identified on the bloodstained bandana found on the balcony outside the bedroom compared to those of the suspect, K.A. When the police were notified of the possible connection between the suspect and the E.S. homicide, they expressed some doubt. Two days later, however, fingerprints on the damaged phone from the victim's bedroom were identified as those of suspect K.A.

The suspect K.A. was subsequently tried and convicted for both the E.S. and A.C. homicides. He was also indicted for the E.S. rape but was not tried because the victim was no longer living.

Both homicides relied heavily on physical evidence. In the A.C. homicide, the bloodstain analysis was used in conjunction with fingerprints of the suspect later developed at the scene as well as his possession of the victim's stolen car and credit cards. In the E.S. homicide, the bloodstain evidence was used as an investigatory tool for the police agency. In both cases, the bloodstain analysis proved to play an integral part in the identification of the victims' assailant.

Table VI. E.S. Homicide—Blood Analysis Results

Specimen	Blood Group	EsD	PGM	GLO I	CA II	ADA	ACP	AK	Hp	Gc	Tf
E.S. blood control	O	2-1	2-1	2-1	1	1	BA	2-1	2	2-1	C
Bedroom											
Telephone cord	—	—	—	—	—	1	BA	2-1	—	—	—
Newspapers	O	2-1	2-1	—	1	1	BA	2-1	2	2-1	C
Knife	O	2-1	2-1	2-1	1	1	BA	2-1	—	—	—
Knitting needle	—	2-1	2-1	2-1	1	1	BA	2-1	—	—	—
Balcony											
Bandana	—	—	—	—	—	1	—	1	2-1	1	CD
K.A. blood control	O	1	1	2-1	1	1	B	1	2-1	1	CD

References

[1] *Technical Manual of the American Association of Blood Banks.* Miller, W.V., Ed.; Philadelphia: J.B. Lippincott Company, 1977; p. 90.
[2] Culliford, B.J. *J. Forensic Sci. Soc.* **1967,** 7, No. 3, 131–33.

[3] Gdowski, F. M.; McCormack, G. L., presented at the Northeastern Association of Forensic Scientists Meeting, Albany, NY, 1982.
[4] Giblett, E.R. *Genetic Markers in Human Blood*. Oxford: Blackwell Scientific Publications, 1969; p. 145.

Glossary of Terms Used in Genetics

If a physical characteristic or trait is **polymorphic,** several distinct types or kinds of features can be specifically inherited. Thus, blood groups are polymorphic, but not height.

The **phenotype** is the physical characteristic produced by the **genotype.** For example, a person with blood group genotype A/A or A/O has phenotype A.

Genetic markers are specifically inherited traits that can be analyzed (for example, proteins in blood).

8

Bloodstain Analysis
Serological and Electrophoretic Techniques

Lawrence Kobilinsky

Forensic serologists play a vital role in our criminal justice system. They will often take advantage of knowledge gained over the past decades in the biochemical, immunological, and biomedical sciences to solve problems related to violent crimes such as rape, homicide, and assault as well as nonviolent legal situations such as disputed paternity.

Serology and Serologists

Serology is a subdiscipline of immunology and has become a great deal more than simply "the study of serum" as its name implies. The serologist utilizes the various methods of analysis of antigens and antibodies and the reactions that occur between them to characterize a sample of biological material, such as blood, or a particular component of that heterogeneous sample.

An *antigen* is a substance that, when introduced into the body of an animal (by injection, inhalation, ingestion, etc.), elicits a specific immune response. The substance in general must be "foreign" to the animal and thus recognized as non-self and also must be of at least a critical size and of a minimal degree of complexity. Examples of antigens are bacteria, viruses, ragweed pollen, dust, drugs, and the constituents of blood. An *antibody*, also termed an *immunoglobulin*, is a blood protein produced by plasma cells in response to stimulation with an antigen. An antibody will combine specifically with its homologous antigen, that same antigen that elicited antibody production.

Many procedures used by the serologist result in the combination of an antigen and an antibody to form an *immune complex*, e.g.,

0784/83/0087$06.00/0
© American Chemical Society

$$\text{antigen} + \text{antibody} \rightleftharpoons \text{antigen–antibody complex}$$

At times, additional techniques are needed to demonstrate that this reaction has taken place. For example, using radioactive tracers such as ^{131}iodine-labeled antigen, one can determine if ^{131}I is present in the newly formed complex (^{131}I-antigen–antibody). Similarly, using fluorescent dye-tagged antibodies, one can detect if a complex has formed by determining if it contains the fluorescent marker.

Often, however, the formation of an immune complex results in a visible reaction such as a precipitation or an agglutination. This reaction will usually require incubation of the reagents at the proper temperature for some period of time. Whether or not a visible reaction will occur depends on the nature of the reaction, the solubility and size of the antigen, the type of antibody, the temperature, the presence of additional serum factors, and other such factors. Precipitation reactions are easily observed after a variable time period following the mixing of the soluble antigen and antibodies. *Agglutination* is the clumping together or flocculation of a particulate antigen in the presence of its homologous antiserum.

Great differences exist between forensic and clinical serology despite the fact that they use similar or identical techniques. Clinical serologists are often asked to determine if two blood samples are compatible so that a transfusion of blood from donor to recipient will not result in potentially dangerous results. They will study the interaction, if any, between antibodies in the host blood and red cell antigens in the donor blood and similarly they will study the interaction, if any, between antibodies in the donor blood and red cell antigens from the host (recipient's) blood. They may be asked to determine if a patient's blood indicates the presence of an infectious agent or they may be asked to determine if a patient has unusually high or low levels of some blood constituent.

Forensic serologists, on the other hand, often deal with samples other than fresh wet blood and are charged with the unique problem of identification and *individualization* (the determination of the source of the material or physical evidence under study). The forensic serologist may be presented with a case involving blood, saliva, semen, or some other physiological fluid or tissue. The sample may be wet or dry and of almost any age. The sample may be fairly "clean" or it may be heavily contaminated with biotic (bacteria, fungi, etc.) or abiotic factors. It may even be a mixture of similar fluids from two separate sources, for example, in a violent crime the possibility exists of encountering blood or bloodstains derived from both the victim and the perpetrator. The sample may have undergone deterioration due to environmental conditions such as humidity, temperature, or even a deliberate attempt by an individual to destroy the physical evidence. The difficulties in analysis are immense, especially when only a limited sample is available. How then can forensic serologists accomplish their mission of identification and individualization? Techniques capable of exploiting differences in individuals must be employed.

Genetics

When an egg cell is fertilized, 23 chromosomes from the mother and 23 from the father join in forming the nucleus of the newly formed cell, the *zygote*. Throughout fetal and postnatal development the genetic constitution of this individual remains more or less fixed (with certain exceptions). *Chromosomes* are the repositories of *genes*, which are the determinants of all physical traits such as height, weight, and eye color. In general, deoxyribonucleic acid (DNA), which constitutes the gene, houses the information required for protein synthesis by the cellular machinery. The information derived from DNA is utilized in the transfer of information (*transcription*) to "messenger" ribonucleic acid (RNA) and this newly formed molecule is then *translated* (supplies the same information but in a different form for copying) by *ribosomes* (cellular elements) into individual structural or regulatory proteins. Two additional types of RNA molecules (transfer and ribosomal RNA) are involved in this process. The result of this process is the formation of a protein that may or may not have enzyme activity depending upon the information originally found in DNA within the transcribed genes. The DNA is a linear array of deoxyribonucleotides whose order ultimately determines the order of the amino acids (building blocks) of all proteins. Thus two molecules of DNA differing in only one or two out of 10,000 nucleotides may result in two different proteins having different characteristics and/or activities.

A *genotype* is the genetic constitution of an individual. A *phenotype* is the physical expression of the genes. A person may have genes for blue eyes and brown eyes (one from each parent) but he or she will be either blue eyed or brown eyed: e.g., genotype, blue/brown; phenotype, blue. Individuals who have two identical genes for a certain trait are called *homozygotes*, and individuals who have two different genes are *heterozygotes*.

Individuals differ not only phenotypically but also on the molecular level. In the case of identical twins who are products of the same zygote, when exposed to similar environmental conditions, individuals generally develop phenotypically identically. Needless to say, there is less similarity between individuals such as fraternal twins, and nonrelated individuals are each genetically unique. This uniqueness is manifested in several ways, including differences in the composition of various body tissues and fluids such as blood. The forensic serologist would like to use the molecular uniqueness of individuals and their various organs and tissues to achieve the goal of identification and individualization.

Blood Chemistry

For illustrative purposes, let us discuss blood chemistry and what kind of information the forensic serologist can learn by analysis of a relatively small amount of bloodstain evidence. *Blood* is a complex mixture of erythrocytes (red cells), leukocytes (white cells), platelets, fibrinogen, and serum. *Serum* is the fluid portion of whole blood after removal of cells and the clotting protein, fibrin. Serum contains (1) electrolytes and metals such as bicarbonate, calcium, chloride, and copper; (2) nutrients such as amino acids and glucose; (3)vitamins; (4) metabolic intermediates such as bile acids, choline, bilirubin, and creatine; (6) hormones; (7) dissolved gases; and (8) proteins. These proteins constitute roughly 7% of plasma and consist of albumin, antibodies, clotting factors, fibrinogen, globulins, complement components, and enzymes. More than 160 antigens, 150 serum proteins, and 250 cellular enzymes have been identified in human blood.[1] These consist of the soluble antigens and those antigens that are present on .the formed elements (erythrocytes, leukocytes, and platelets).

Some erythrocyte antigens are very common in the population and are classified as factors of the primary blood groups such as ABO, MNSs, Rh, Le, and Lu. Approximately 14 systems are considered as primary. Some erythrocyte antigens are less common and are therefore classified as factors of secondary systems (*see* Table I). Some antigens (A and B) are present on all three types of formed elements of whole blood, while others are unique to leukocytes and platelets (histocompatibility locus antigens),

Table I. Erythrocyte Blood Grouping Systems

Primary		Secondary	
ABO	Auberger	En	Ot
MNSs	August	Gerbich	Raddon
Rh	Batly	Briffith	Radin
Lewis (Le)	Becker	Good	Rm
Lutheran (Lu)	Biles	Heibel	Stobo
P	Bishop	Ho	Swann
Kell	$Bg^aBg^bBg^c$	Ht^a	Torkilden
I	Box	Jn^a	Traversu
Duffy	Cavaliere	Kamhuber	Vel
Kidd	Chido	Lan	Ven
Diego	Chr^a	Levay	Webb
Dombrock	Cost	Ls^a	Wright
Xg	Dp	Marriot	Wolfshag
Yt	El	Orris	

and others are found exclusively associated with erythrocytes. Over 250 proteins have been found in or on erythrocytes with approximately 95% (by weight) being hemoglobin. Most erythrocyte proteins are water soluble. There are also numerous proteins found within serum. Some serum proteins are important in normal blood function and are found in very low levels such as prostatic acid phosphatase, alkaline phosphatase, lipase, or alcohol dehydrogenase. In addition, serum contains hormones such as insulin, glucagon, growth hormone, and erythropoietin, which are all protein in nature.

Analysis of Evidential Specimens

The forensic approach to serology follows an orderly format. When confronted with physical evidence that appears to be a bloodstain, the forensic serologist first determines if the stain is really blood. If it is then found to be blood, he or she proceeds to determine the species of origin. The identification of blood may be based on histological or serological analysis or by chemical testing.

Preliminary Tests

Preliminary, presumptive tests are based on color development that indicates the possibility that the stain is blood. Such tests determine if heme or a heme derivative is present and are highly sensitive (to 1 part per million), but lack absolute specificity because many substances other than blood can cause a positive reaction. The presumptive test is followed by a specific assay that is generally far less sensitive. These specific tests include crystal and spectrophotometric tests.

The three most commonly employed *crystal tests* include the Teichmann (hematin), Takayama (hemochromogen), and acetone–chlor–hemin tests. Each of these is rapid and specific. The microscopic identification of the characteristic crystal formed in each test is considered proof beyond doubt that the material is blood.

Spectrophotometric analysis is a method of chemical analysis based on the fact that different substances absorb light rays at specific wavelengths. The absorption spectrum of a substance is a graphic display of its degree of absorption of light as a function of wavelength. A sample is irradiated, and the amount of light that the sample then transmits is measured. Different concentrations of the same sample transmit different amounts of light. In forensic bloodstain analysis, the spectrophotometric method is based on determining if the material in question has an absorption spectrum characteristic of heme or its derivatives.[1] This method yields specific results only if multiple absorption spectra are determined following specific chemical treatments of the sample. Some substances from plant matter can cause great difficulties in interpretation because their spectra often appear similar to hemoglobin or one of its derivatives. For this reason, multiple absorption spectra should be studied before reaching any conclusions.

Species of origin is then determined by serological analysis utilizing a specific antiserum. This reagent will detect the presence of a species-specific antigen within the sample. The exact method employed by the serologist depends on what type of information is required (qualitative or quantitative); however, an immunologic method is only as good as the antiserum used in the assay and therefore care must be exercized in verifying the specificity of the antiserum. Species determination can also be performed economically and reliably using an antiserum directed against human hemoglobin because this technique can identify the stain as human blood in a single analysis.

Individualization

Following species determination, the serologist can now attempt individualization, provided that there is sufficient sample for testing and that it is not contaminated or deteriorated. Initially, the sample is typed for the primary blood group antigens such as ABO and Rh. In dried blood, unlike fresh wet blood, the erythrocytes have lysed (split open), and this presents further problems for blood typing because agglutination of membrane fragments, unlike whole erythrocytes, cannot be easily observed. Furthermore, some antigens (MN system) are unstable in the dried state, which makes it most difficult if not impossible to type bloodstains accurately. Most forensic laboratories seldom do more than type ABO and Rh from among the list of cellular antigens in Table I when only dried blood is available. Fortunately, many serum proteins such as haptoglobin (Hp), group specific component (Gc), and transferrin (Tf) remain stable in dried stains and can be safely analyzed. These serum proteins together with the cellular proteins constitute a group of blood components of great value to the serologist. The techniques used to analyze bloodstains for these proteins will be discussed in the next section.

Most recently much research has been conducted to analyze bloodstains for sex (by determining the presence of the X and Y chromosomes in leukocytes and other nucleated cells or by determining the ratio of the sex hormones, testosterone to estradiol in the stain), age (by studying the kinetics of the conversion of one isozyme to another or the conversion of an active form to an inactive form), and race. Ethnicity can sometimes be determined by analyzing the following cellular and/or serum proteins: hemoglobin, peptidase A, glucose-6-phosphate dehydrogenase, carbonic anhydrase II, and Gm. Lee[1] gives a complete review of bloodstain typing procedures.

Electrophoretic Analysis

For forensic purposes the serologist would like to have available a technique that is rapid, reliable, sensitive, and economical and that would allow analysis of wet or dry blood samples to determine the factors present that make that particular sample unique—*individualization*. All human individuals are in fact genetically and/or phenotypically unique. This includes identical twins who develop different fingerprint patterns during embryological growth and who may also have differences in circulating antibodies due to differences in exposure to antigenic materials. Similarly, an individual's blood is unique. To study the constituents of blood that confer uniqueness one has to separate the various protein components of the mixture. Blood proteins can be separated based on (1) molecular size, by using procedures such as dialysis, ultrafiltration, density gradient centrifugation, or gel filtration; (2) solubility, by using methods such as isoelectric precipitation, salting in or salting out, or solvent fractionation; and (3) electric charge differences, by using affinity chromatography, selective adsorption, ion exchange chromatography, or electrophoretic methods.

The most rapid, economical, and accurate technique that can be adapted to routine forensic casework is electrophoresis. Indeed clinical laboratories include serum electrophoresis as a routine screening procedure for patient sera.[2] The technique of electrophoresis was first developed by A. Tiselius in Sweden in the 1930s. Since that time many modifications of the original ingenious concept have made electrophoresis one of the most commonly used high-resolution procedures available.

Proteins have a broad range of molecular weights, ranging from 5,000 to more than 1,000,000 daltons; most proteins contain from 100 to 300 amino acids. Proteins have three-dimensional conformation (structure); some are globular, others are fibrous, and still others have a combination of both characteristics. The biological properties of proteins include sensitivity to temperature and pH (acidity). Proteins can be classified as structural or enzymatic depending on their function and their types and number of amino acids. The net charge on a protein is determined by its constituent amino acids and by the pH at which it is placed. Because unlike charges attract one another, a negatively charged protein in an electric field migrates toward the anode (+ electrode), and a positively charged protein migrates toward the cathode (− electrode). Proteins in a mixture have different mobilities as a result of differences in charge-to-mass ratio at particular pH values and thus will separate from one another.

In zone electrophoresis, the mixture of substances to be separated is placed in a narrow band (the origin) between the electrodes. As the electric field is applied, the various proteins, each of which has different mobility, move away from each other and thus this procedure results in a separation.

To avoid or at least slow the rate of diffusion of the various separated proteins, a solid or semisolid stabilizing support medium is used. After the separation has been achieved, the proteins are either fixed in position with a fixative or by immunofixation using monospecific antiserum or else the position of the protein is determined by specific enzyme staining and generally photographed for court presentation.

Interpreting Results

In practice, if a serologist examines two samples, one from a crime scene and known not to belong to the victim and a second from a suspect, and if these two samples are phenotypically different based, for example, on the electrophoretic analysis of PGM, the serologist could thereby exclude the suspect. However, if the banding patterns are identical, further testing is necessary to draw any conclusion. When comparing two samples, only one phenotypic difference found as a result of a multitude of electrophoretic analyses is sufficient to rule out the suspect. If, on the other hand, all enzyme tests indicate similar phenotype patterns, a statistical statement can be made as to the probability that the two stains have a common origin.

For example, if a person's blood is typed in two systems, A and B, and the isozyme pattern of A and that of B reflect population frequencies of 10 and 30%, respectively, then it can be stated that the blood under study is relatively rare and is expected to occur in 3% of the population (10% × 30%), assuming that the two systems, A and B, are inherited independently of each other (either found on different chromosomes or if located on the same chromosome, then not closely linked). If a third test indicates a phenotype that is present in 20% of the population, then this blood sample is even rarer, being present in only 0.6% of the population (10% × 30% × 20%). Under certain circumstances, that is, where very rare phenotypes are observed, one can narrow the statistics for that blood sample to one in one million or more. This is clearly of great importance and demonstrates the usefulness of the technique.

To quote Laudel, Grunbaum and Kirk, "It is a basic assumption of the criminalist that Nature never repeats herself. . . . it should be possible to establish the individuality of every object. This is the basis upon which criminalistics rests."[3] Although serological analysis has not yet reached the capabilities of fingerprint analysis in achieving the goal of individualization and identification, the electrophoretic technique certainly allows the forensic serologist to approach that goal more closely than has ever before been possible.

References

[1] Lee, H. C. In "Forensic Science Handbook"; Saferstein, R., Ed.; Englewood Cliffs, NJ: Prentice Hall, 1981; pp. 267–337.
[2] Wolf, P. L. "Interpretation of Electrophoretic Patterns of Protein and Isoenzymes," 1st ed.; New York: Masson, 1982.
[3] Laudel, A. F.; Grunbaum, B. W.; Kirk, P. L. J. Forensic Sci. 1963, 10, 57–64.

9

Does the Crime Laboratory Have the Answers?
Four Cities Compared

Joseph L. Peterson

Most treatments of forensic science in textbooks and scientific journals discuss laboratory findings and interpretations from the individual case perspective. The technical literature customarily focuses on the scientific methods that may be employed to examine various types of physical evidence and the kinds of information that may be derived from these different analytical approaches. The popular forensic science literature highlights those few outstanding cases where the crime laboratory was successful in solving major crimes and convicting the rightful offender. Of late, too, there is a growing number of case histories in the literature in which the analysis of physical evidence or the testimony of experts in court went awry, demonstrating the dangers associated with improperly collected or examined evidence and expert court testimony offered by less than qualified experts.

Almost no published research examines the process of evidence retrieval, analysis, and interpretation from a case aggregate level. A review of the forensic science literature tells us little about the kinds of physical evidence collected in typical criminal investigations, or the results and types of information that are routinely derived from its laboratory analysis. Putting extraordinary cases aside, we have few insights into the benefits derived from the analysis of physical evidence by forensic scientists in publicly supported crime laboratories. Such studies, though, are critical to our understanding of forensic science and the possible benefits that can be anticipated from the examination of evidence in the future.

Forensic Evidence and the Police Research Study

The purpose of this chapter is to discuss data that were gathered in a recently completed two-year study of forensic evidence funded by the

0784/83/0097$08.50/0

National Institute of Justice and the Forensic Sciences Foundation. This project had the goal of examining the uses and effects of scientifically analyzed evidence within the context of police investigations. Four jurisdictions were selected for this research, one small (Peoria, Ill.), two medium-sized (Kansas City, Mo., and Oakland, Calif.), and one large (Chicago, Ill.), to represent various sized jurisdictions and approaches to physical evidence utilization. Appendix A contains a summary of background information about the police and scientific resources in each of these jurisdictions.

The basic data collection approach used in this study was a random selection of completed investigations from the files of the police agencies and crime laboratories. Approximately 1,600 case files were reviewed in which physical evidence had been collected and analyzed. This review started with the report of each of these offenses to the police, the investigation strategies used by the police agency, the physical evidence collected and examined, the manner in which the police disposed of the case, and the court disposition of persons arrested during the course of these investigations.

In addition, comparison samples of 1,000 robbery, aggravated assault, and burglary cases were randomly selected in which physical evidence was *not* collected or examined. This sampling design enabled us to compare the police and prosecution outcomes of similar cases (by crime type), while controlling for the presence or absence of various types of physical evidence. The final report on this project, *Forensic Evidence and the Police* (Peterson et al., 1982), fully discusses the data collected and study results.

This chapter will focus on those approximately 1,600 cases where physical evidence was collected and analyzed in the laboratory, centering primarily on the results of scientific testing of various types of evidence. Table I summarizes the cases (by crime classification) that were included in this sample.

Evidence Utilization Process

The flow chart in Figure 1 depicts the various decision points and decision makers who are central to understanding the collection and analysis of physical evidence. Two fundamental tracks are depicted in this chart, the physical evidence path and the police investigation path; these occur concurrently and with considerable interchange among the principal parties involved. Our primary concern is with the decision process that controls the flow of physical evidence and the information derived from it.

The objectives of this chapter, then, are to provide

- a summary of crime incident variables associated with what physical evidence is gathered from scenes of crimes;

Table I. Total Crimes in Physical Evidence Sample

Crime Type	Peoria	Chicago	Kansas City	Oakland	Total
Homicide	29	72	51	71	223
Other deaths	21	7	0	1	29
Rape/sex offenses	53	53	49	70	225
Robbery	17	36	57	39	149
Aggravated assault	66	62	49	34	211
Burglary/property	55	80	52	42	229
Arson	2	40	44	0	86
Weapons related	39	24	0	4	67
Drugs	52	54	46	73	225
Fraud/forgery	0	13	55	0	68
Other	48	15	1	15	79
Total	382	456	404	349	1,591

- the primary reasons evidence is gathered and submitted for analysis;
- the ratio of evidence submitted for analysis to that actually examined;
- the results of laboratory testing by crime and evidence type; and
- a discussion of fingerprint results derived from a special sampling of cases where *only* latent fingerprints are gathered from the scenes of crimes.

Incident Variables Associated with the Number of Evidence Categories Collected

An examination of cases in the study revealed, first, a wide range of evidence types gathered in various types of offenses. The quantity of evidence (number of discrete categories of evidence) collected is associated with various circumstances surrounding the crime, factors termed "incident variables."

Table II identifies those incident variables, in personal and property crimes, that have a positive association with the number of categories of evidence collected. These relationships are distinguished by the type of crime committed (personal or property) because the direction and significance of the relationships are sometimes different.

The chi square test of significance legend at the bottom of the table indicates the *strength* of the relationship between the various independent

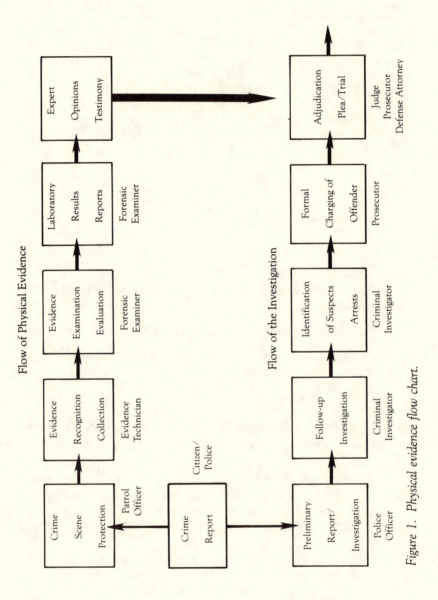

Figure 1. Physical evidence flow chart.

Table II. Incident Variables That Have a Positive Association with the Number of Evidence Categories Collected

Incident Variable	Peoria	Chicago	Kansas City	Oakland
More evidence is collected:				
• in personal rather than property offenses.	N.S.	•••	•••	•••
• as the injury sustained by the victim in personal offenses increases.	•••	•••	•••	•••
• when the offender has a physical interaction with the victim and/or scene.	•••	•••	•••	•••
• from residential scenes in personal crimes.	••	N.S.	N.S.	•
• from residential scenes in property crimes.	(−) ••	•	•••	N.S.
• when the suspect is *not* identified or in custody in personal crimes.	•••	N.S.	•	•••
• when the suspect *is* identified or in custody in property crimes.	N.S.	••	••	N.S.
• when witnesses are *not* present in personal crimes.	•••	•••	•••	•••
• when detectives/supervisors are present at personal crime scenes.	•••	•••	•••	•••
• when detectives/supervisors are present at property crime scenes.	••	••	•••	N.A.

Key: N.S., not significant; N.A., not applicable; (−), negative association; and the chi square significance is •, $p < 0.05$; ••, $p < 0.01$; •••, $p < 0.001$.

variables and the dependent variable, the number of evidence categories collected. A relationship that is found to be significant means that the null hypothesis (complete *independence* between the independent and dependent variables) is rejected. In other words, there *is* a relationship between the two variables. The p factor (< 0.05, < 0.01, or < 0.001) gives the approximate probability that we would find such an association by chance (when, in fact, the two variables are truly independent of one another) and is less than 5 in 100 (*); 1 in 100 (**); or 1 in 1000 (***).

Crime Classification

In all cities except Peoria, significantly more evidence is gathered in crimes against persons than in crimes against property. In Oakland, for example, four or more evidence categories are collected in 70% of the personal crimes, while in just 14% of property offenses. In 36% of the property offenses only a single evidence category is collected, versus only 9% of the personal crimes. In Peoria, the quantity of physical evidence collected in property crimes is not significantly different from the number of categories collected in personal crimes. None of these single evidence category cases involves only fingerprints. These cases are considered as a separate category and are discussed later.

Personal Injury

In personal crimes (murder, rape, assault, and robbery) in all jurisdictions, the amount of evidence collected is highly associated with the seriousness of physical injury suffered by the victim. When the victim receives either a minor injury not requiring medical treatment or no injury at all, only one or two categories of evidence are collected in a majority of the cases. But as the degree of injury becomes more severe, the quantity of evidence collected steadily increases. Table III illustrates this steady progression in Kansas City. The exception is Oakland where high quantities of evidence are collected in even the least serious offenses.

This relationship is probably due to the following: (1) the amount of evidence generated during the commission of the crime (more violent crimes produce more evidence); and (2) the added motivation of technicians to collect evidence when investigating more serious offenses.

Interaction Between Offender and Scene and/or Victim

Not all personal crimes involve struggle or physical contact between the offender and the victim. Robberies frequently do not involve physical interaction between offender and victim. In such cases, one would not expect to find or recover the same quantity of physical evidence as in cases where there is such interaction. The data from all the cities support

Table III. Crimes Against Persons in Kansas City; Extent of Injury by Number of Physical Evidence Categories Collected

	Number of Evidence Categories Collected				Row Total
Personal Injury	1	2	3	4 or more	
None/minor	27	29	20	24	34
Moderate	0	10	13	77	19
Serious	11	27	35	27	22
Fatal	0	0	4	96	25
Column total	11	18	18	53	100

NOTE: Both column total and row total percentages add to 100%. The individual cell entries in the table are row percentages and, accordingly, should be read horizontally. For example, "27% of the crimes against persons, where injuries to the victim were classified as 'none or minor', involved only one category of evidence being collected."

The chi square significance is $p < 0.001$; $N = 207$.

this theory; statistically significant associations ($p < 0.001$) were found between interaction and number of evidence categories collected. For example, in Peoria, four or more categories of evidence are collected in 52% of the cases with a physical interaction, but in only 6% of the incidents without such an interchange.

Location of the Offense

In personal crimes, more evidence is usually gathered from residential crime scenes than from commercial scenes or incidents occurring on the street or out-of-doors. This relationship is strongest in Peoria; the weakest relationship is in Chicago, where no association is found. The results for property crimes are not consistent across all the cities. Peoria evidence technicians tend to gather more evidence at nonresidential locations, but the opposite is true in the other cities.

Status of the Identification of the Suspect

The following interesting relationship is consistent in three of the four study jurisdictions. Basically, more physical evidence categories are collected in personal crimes when the *least* information about the identity or whereabouts of the suspect is available. The fewest categories of evidence are gathered when a suspect is in custody. This pattern of collecting less physical evidence when a suspect is in custody is understandable because such cases practically always have a witness to corroborate the suspect's involvement. This situation reduces the need for physical evidence to link a suspect with the crime. Technicians make an extra effort when suspects are not in custody or identified in some fashion.

Chicago is the only exception to this pattern. The amount of evidence collected appears to be unrelated to the status of the identification of the suspect. Chicago also generally collects the fewest categories of evidence per crime of all the jurisdictions.

The opposite trend is true for crimes against property; more evidence is collected in offenses with a suspect in custody, at least in Chicago and Kansas City. Given the low probability of solving property offenses when a suspect is neither in custody nor identified at the beginning of the investigation, technicians may have learned through experience that there is little payoff in collecting many categories of evidence in such cases. When a suspect is in custody, though, the technician is presented with an opportunity to corroborate that suspect's involvement through physical evidence (e.g., to place a suspect apprehended on the street *inside* a dwelling through fingerprints or trace evidence). This corroboration may be particularly important in burglary/property crimes because witnesses are rarely present.

Witnesses to the Crime

As in the preceding variable, in crimes against persons more evidence is usually collected when there are no eyewitnesses to the crime. In property offenses, as with the suspect identification variable, more evidence is collected when there are one or more witnesses. This finding indicates that more evidence is collected in property crimes when investigators have good leads to start with, as is also evident with the suspect variable.

Police Personnel at the Scene

The relationship between evidence gathered and the presence of detectives and other supervisory personnel at the crime scene was also examined. The data support the theory that technicians collect more evidence when these personnel are present. This significant relationship suggests that technicians are subject to the same pressures from higher police authorities that other personnel in the department feel, and will perform a more exhaustive search in their presence. This relationship is also probably affected by the fact that supervisors will more likely be present at the more serious offenses. This condition has already been shown to be associated with more evidence being collected.

Collecting Agent

When the types of police personnel collecting evidence in the case are cross-tabulated by the number of evidence categories collected, patrol officers are shown to have a decreasing likelihood of collecting multiple forms of evidence. Evidence technicians, detectives, and medical personnel are the primary collectors of multiple categories of evidence. Table IV illustrates this relationship for personal crimes in Kansas City. A patrol officer is a collecting agent in only 17% of the cases where four or more categories of evidence are collected. Table V shows the percentage of time in which the various types of personnel are collectors of evidence in cases where four or more categories of evidence are collected.

Table IV. Crimes Against Persons in Kansas City; Number of Physical Evidence Categories Collected by Collecting Agent

Number of Evidence Categories	Personnel Collecting Physical Evidence				Row Total
	Police Officer	Detective Supervisor	Evidence Specialist	Medical Personnel	
1	71*	8	25	8	12
2	24	32	68	16	18
3	16	49	89	19	18
4 or more	17	66	94	74	52
Column total	25	50	81	46	100

NOTE: Cell entries are row percentages. $N = 207$.

* This value should be read, "A police officer was one of the collecting agents in 71% of the cases in which one category (blood, trace, etc.) of evidence was collected."

Table V. Percentage of the Time Various Police Personnel Are Collecting Agents in Cases in Which Four or More Categories of Evidence Are Collected

Collection Agent	Peoria	Chicago	Kansas City	Oakland
Police officer	20	32	17	42
Detective/supervisor	86	81	66	66
Evidence specialist (technicians, criminalists)	93	79	94	68
Medical personnel (medical examiner, doctors, nurses)	77	80	74	69

Categories of Physical Evidence Collected

Table VI enumerates the top five evidence categories collected in the combined total of all homicides, rapes, robberies, assaults, and bur-

Table VI. Leading Evidence Categories Collected in All Crimes Combined

Peoria (N = 241)	Chicago (N = 310)	Kansas City (N = 258)	Oakland (N = 257)
Firearms (52%)	Firearms (40%)	Fingerprints (63%)	Blood (52%)
Blood (32%)	Blood (38%)	Firearms (29%)	Fingerprints (49%)
Fingerprints (28%)	Fingerprints (23%)	Blood (21%)	Firearms (47%)
	Questioned documents (13%)		
Hair (23%)		Hair (18%)	Hair (24%)
Semen (14%)	Semen (13%)	Fire related (14%)	Semen (23%)

glaries in the sample. These further observations are in order:

- Biological fluids and firearms dominate as evidence collected in crimes of violence.
- Fingerprints, trace evidence, and toolmarks are the leading evidence categories collected in property crimes.
- Oakland has the highest percentage of personal crimes with blood evidence and firearms; Peoria has the lowest percentage of cases with blood evidence.
- Chicago has the lowest percentage of crime laboratory analyzed cases with fingerprints and trace evidence, while Kansas City has the highest percentage of cases with those same evidentiary items.

Seriousness of the Offense and Evidence Collected

As the seriousness of the personal offense increases, so does the likelihood that biological fluids will be collected. The same relationship is particularly strong in the areas of trace evidence and fingerprints, as well. No clear relationships are seen between the dollar loss sustained in a property offense and the types of evidence collected.

Interaction and Evidence Collected

Interaction between the offender and victim predictably generates not only more biological evidence, but also more trace evidence and fingerprints. The only countertrend here is found with firearms because firearms are likely to be submitted in incidents in which physical interaction has *not* occurred. In such cases a firearm is used as the weapon to intimidate or, possibly, shoot a victim, but the offender does not personally engage in an altercation with the victim. Here firearms may also constitute the source of some other type of evidence, such as fingerprints or bloodstains, which may be deposited on a weapon.

Biological and trace evidence are found only in those property crimes involving an interaction between the offender and the crime scene. On the other hand, fingerprints and tools are collected more frequently in offenses in which no appreciable interaction between the offender and scene has taken place.

Reasons for Submitting Evidence for Laboratory Analysis

Table VII summarizes the various reasons that evidence is submitted to the laboratories for analysis. The N values in this table refer to the various reasons that evidence is submitted in a case. Because individual

Table VII. Reasons for Submission of Evidence

Reasons	Peoria N = 862	Chicago N = 1139	Kansas City N = 1139	Oakland N = 715
Element	8	9	10	9
Associative	62	44	52	63
offender/scene	35	28	55	32
offender/victim	23	9	8	24
firearm related	34	43	24	38
victim/scene	4	8	12	5
tools	2	1	1	—
documents	—	9	—	—
Reconstruct	13	32	32	13
Corroborate	4	6	5	10
Operability (firearms)	13	9	1	5
Total	100	100	100	100

NOTE: Results are in percent.

cases often involve more than a single category of evidence and because a category of evidence may be submitted for more than one reason, the N values (reasons) are greater than the number of cases sampled in each jurisdiction.

Element of the Crime

An examination of the cases sampled shows that evidence is submitted for the purpose of establishing an element of an offense from 8 to 10% of the time. Drug and narcotic offenses are not included in this accounting. However, cases in which drugs are submitted as evidence incidental to the major crime category are included, such as cases in which drugs are found in the automobile of a robbery suspect. Therefore, rape and arson are the two primary crime categories in which evidence is submitted to establish an element of the crime. In such cases, suspected seminal fluid and flammable substances are submitted for reasons of identification.

Associative Evidence

The primary reason evidence is submitted in the cases sampled in all jurisdictions is to associate persons, instruments of the crime (firearms, other weapons, and tools), and locations where offenses occur. Peoria (62%) and Oakland (63%) have the greatest percentage of evidence submitted for this purpose, while Kansas City (52%) and Chicago (44%) have evidence submitted for this purpose to a lesser degree.

Within the association category, the submissions in Peoria and Kansas City are primarily intended to associate offenders with the scenes of crimes. In Chicago and Oakland, the majority of the submissions are related to firearms and are intended to associate these weapons with their owners, with the offenders, or with the victim of the crime. A substantial difference is seen among the study cities in the fraction of submissions where the intent is to associate the offender with the victim of the crime. Approximately one-quarter of this associative evidence in Peoria and Oakland has the objective of linking an offender with a *victim,* while less than 10% of the associative evidence in Chicago and Kansas City is submitted for that purpose. This result is, in part, a reflection of the higher percentage (80%) of personal crimes in the Peoria and Oakland samples, compared with Chicago and Kansas City where only about 70 and 60%, respectively, of the cases are personal crimes.

Reconstruction

About two and a half times more cases are submitted in the Chicago and Kansas City samples where one of the primary reasons for submission is reconstruction. This situation reflects the fact that a higher percentage of cases examined in Chicago and Kansas City lack standards. For example, bloodstain evidence from a crime scene is examined, but no blood sample is submitted from a known source (i.e., the victim or offender). In such cases, the examination can provide information about the blood type of the individual who shed the blood, but cannot associate it with anyone.

Corroboration

Evidence is submitted between 4 and 10% of the time to test the statements of witnesses and victims and the alibis of suspects. This reason for submitting evidence is common in cases of rape where testing the evidence taken from the victim would support or refute the statements she has given the police.

Operability/Open Case File Check

A substantial volume of firearms evidence in Peoria and Chicago has been examined for the purpose of checking the operation of the weapon and comparing the weapon against open case files to see if the gun may have been involved in previous crimes. Almost 10% of the Peoria caseload sample involves unlawful use of weapons. In order to prosecute, the laboratory has to verify that the gun is in operating condition.

Ratio of Evidence Examined to Evidence Collected

Table VIII details the average number of discrete evidence categories collected and examined by type of offense in the four cities. The fraction in the columns beneath each city divides the average number of evidence categories examined per case by the average number of categories collected per case. Examiners in Peoria examine the highest percentage of categories collected in four crime categories. Oakland examines the lowest percentage of evidence categories collected in all five primary offenses. In homicide, Oakland evidence technicians collect an average of 6.3 categories of evidence per investigation, but the laboratory only examines an average of 1.8 categories per case. The Oakland laboratory examines, on the average, 1.4 categories of evidence in rape cases (the lowest of all the cities) but technicians gather 5.2 categories per case (the highest of all the cities, along with Kansas City).

In all cities, except for Kansas City, the highest ratio of evidence examined to evidence collected is in burglary/property offenses. The lowest ratio of evidence examined/collected is in homicides. This result is undoubtedly related to the higher than average quantities of evidence collected in those very serious offenses. It appears, though, that laboratories screen out much of this evidence from their examination procedures.

Table VIII. Percent of Physical Evidence Categories Collected Which Are Examined by Crime Type

Crime Classification	Peoria		Chicago		Kansas City		Oakland	
	N*	Percent	N	Percent	N	Percent	N	Percent
Homicide	$\frac{2.2}{4.3}$	51	$\frac{2.0}{4.0}$	50	$\frac{3.3}{5.8}$	57	$\frac{1.8}{6.3}$	29
Sex crimes	$\frac{2.4}{3.2}$	75	$\frac{1.8}{2.8}$	64	$\frac{2.7}{5.2}$	52	$\frac{1.4}{5.2}$	27
Robbery	$\frac{1.4}{2.0}$	70	$\frac{1.5}{2.2}$	68	$\frac{1.5}{3.0}$	50	$\frac{1.3}{3.4}$	38
Assault	$\frac{1.4}{1.9}$	74	$\frac{1.3}{2.1}$	62	$\frac{1.3}{1.9}$	68	$\frac{1.1}{3.0}$	37
Burglary	$\frac{1.4}{1.7}$	82	$\frac{1.1}{1.5}$	73	$\frac{1.5}{3.0}$	50	$\frac{1.1}{1.7}$	65
Arson	—	—	$\frac{1.1}{2.2}$	50	$\frac{1.3}{2.3}$	57	—	—

*Fraction represents mean number of evidence categories examined divided by the mean number of evidence categories collected.

Laboratory Results

Laboratory Results by Crime Classification

Table IX exhibits the results of laboratory testing in each jurisdiction by personal and property crimes. The N values in the table refer to the number of evidence categories submitted and analyzed by the laboratory in the sample of cases from each jurisdiction. The percentage totals for each crime classification exceed 100% because the data collection instrument used in the study recorded up to three results for each major category of evidence collected. Although an infrequent occurrence, a case might involve several different blood samples submitted from various locations at different crime scenes. In such a case, one sample might prove inconclusive, while another is typed and associated with a suspect. However, most cases have a single result.

If the examination results in the identification of the evidence (e.g., the stain is blood, the liquid is flammable), or a classification (the stain is Type A human blood, the flammable liquid is gasoline), it is included in the identify/classify category. Chicago has the highest percentage of results in both the personal and property crime category when the results are so classified.

Initially, most types of evidence are identified or classified even if the evidence is compared subsequently with a standard, thus yielding a conclusion concerning the origin of the evidence. If a blood sample is first grouped and then compared with blood that has been grouped from another source, and a statement of common origin results (in the above example, the two samples *possibly* have a common origin), both the "identify/classify" and the "common origin" results are noted.

The second row notes negative identifications. For example the evidence is determined *not* to be the substance it was thought to be upon submission. The primary evidence forms here would be substances suspected to be seminal fluid, flammable liquids, controlled substances, and bloodstains. A small percentage of the time these substances turn out to be something else, such as a prescription drug rather than a controlled substance or the laboratory is unable to detect the presence of the substance due to the small sample size or contamination/deterioration of the sample.

The percentage of results that possibly, probably, or conclusively link evidence with a standard (and thereby its source) are all categorized under the common origin heading. Results from the examination of cases in Peoria are in the common origin category more often than the cases from the other cities. Forty-four percent of the results in personal crimes and fifty-four percent of the results in property crimes are of the common origin category. Chicago has the lowest percentage of results classified in the common origin category; 21% of the results are from personal crimes and

Table IX. Results Derived from the Laboratory Examination of Physical Evidence

Laboratory Result	Peoria		Chicago		Kansas City		Oakland	
	Pers. (N = 421)	Prop. (N = 97)	Pers. (N = 411)	Prop. (N = 123)	Pers. (N = 431)	Prop. (N = 161)	Pers. (N = 332)	Prop. (N = 48)
Identify/classify	36	20	58	49	41	29	42	17
Negative identification	5	2	5	11	3	9	8	0
Common origin	44	54	21	5	29	12	35	27
Different origin	5	12	1	2	7	7	16	31
Reconstruct	5	0	10	2	11	14	6	2
Inconclusive	25	20	20	38	24	49	13	25

NOTE: The N value in this table refers to the total number of categories of evidence analyzed by the laboratory of the cases included in the study sample. All values are given in percent.

5% of the results are from property crimes. Kansas City and Oakland are comparable in the personal crime category results, but Oakland has about twice the percentage of common origin results in the property crime category as Kansas City. The sample sizes in these property offense comparisons are noteworthy; the two cities with the lowest percentage of common origin results process the greatest number of cases, by a factor of two to three. This finding suggests that Chicago and Kansas City are not as discriminating in the property cases they choose to examine, while Peoria and Oakland reserve their examinations for cases in which both evidence and standards are supplied.

The Oakland laboratory has the highest percentage of laboratory reports that concluded that two items of evidence did *not* have a common origin. Oakland examiners are more explicit in their laboratory reports about the failure of two items to match with one another and, thereby, indicate that they do not share a common source. The other laboratories tend to declare inconclusive results in such cases. Different origin results constitute valuable information, for they may demonstrate to investigators that they are pursuing the wrong suspect or are operating under a faulty hypothesis as to how the crime occurred. In all cities, evidence submitted in property crimes is more likely to result in a different origin result than that submitted in a personal crime.

Inconclusive results develop when the laboratory is unable to make any type of a definitive statement. Even though Peoria has one of the highest rates of common origin results, it also has the highest rate of inconclusive results in personal crimes. Kansas City has the highest rate of inconclusive results in property crimes with almost half the results falling into this category.

Laboratory Results by Evidence Category

Tables X through XIII summarize the results of laboratory testing for each jurisdiction by evidence category. The N values correspond to the number of times various categories of evidence are submitted in personal and property crimes. Given the infrequency with which some evidence categories appear in certain crime categories, percentages are given only when the N value is five or more cases.

The percent of time that bloodstain evidence results in a conclusion of common origin (probable or possible) ranges from a high of 40% of submissions in Oakland to a low of 6% in Chicago. Blood is rarely present in property crimes in Peoria, Kansas City, and Oakland. But, in Chicago ($N = 25$), blood links an offender with a scene or victim 8% of the time.

Chicago has the highest rate (79%) of positive identifications of suspected semen evidence in rape or other sex-related crimes. The rate

Table X. Peoria—Laboratory Results by Evidence Category and Crime Classification

Evidence Category	Crime Type	N*	Laboratory Results (%)					
			Identification	Negative Identification	Common Origin	Different Origin	Reconstructive	Inconclusive
Blood	Pers.	86	90	2	29	1	1	12
	Prop.	4	—	—	—	—	—	—
Semen	Pers.	43	67	32	5	0	0	2
	Prop.	0	—	—	—	—	—	—
Hair	Pers.	56	20	0	32	20	2	30
	Prop.	1	—	—	—	—	—	—
Firearms	Pers.	149	7	0	62	1	14	49
	Prop.	14	36	0	21	0	0	64
Toolmarks	Pers.	3	—	—	—	—	—	—
	Prop.	22	9	0	82	9	0	14
Prints	Pers.	42	2	0	81	14	0	2
	Prop.	15	0	0	53	13	0	33
Trace/transfer	Pers.	14	14	0	57	21	0	14
	Prop.	21	0	0	62	33	0	0
Drugs	Pers.	25	76	24	0	0	0	0
	Prop.	11	82	18	0	0	0	0
Flammables/explosives	Pers.	3	—	—	—	—	—	—
	Prop.	0	—	—	—	—	—	—
Impressions/patterns	Pers.	10	10	0	60	0	40	10
	Prop.	9	0	0	78	11	0	11

*Values where $N < 5$ cases are not computed.

Table XI. Chicago—Laboratory Results by Evidence Category and Crime Classification

Evidence Category	Crime Type	N*	Laboratory Results (%)					
			Identification	Negative Identification	Common Origin	Different Origin	Reconstructive	Inconclusive
Blood	Pers.	139	95	4	14	0	1	1
	Prop.	25	96	4	8	0	0	0
Semen	Pers.	48	79	17	0	0	0	4
	Prop.	0	—	—	—	—	—	—
Hair	Pers.	19	79	0	11	11	0	16
	Prop.	0	—	—	—	—	—	—
Firearms	Pers.	157	26	0	34	2	25	37
	Prop.	14	7	0	7	7	7	79
Toolmarks	Pers.	5	40	0	0	0	0	60
	Prop.	21	67	0	0	0	5	29
Prints	Pers.	23	0	0	39	4	0	57
	Prop.	23	0	0	13	0	0	87
Trace/transfer	Pers.	2	—	—	—	—	—	—
	Prop.	1	—	—	—	—	—	—
Drugs	Pers.	3	—	—	—	—	—	—
	Prop.	0	—	—	—	—	—	—
Flammables/explosives	Pers.	13	46	54	0	0	0	15
	Prop.	34	56	35	0	0	0	24
Impressions/patterns	Pers.	2	—	—	—	—	—	—
	Prop.	3	—	—	—	—	—	—

*Values where N < 5 cases are not computed.

Table XII. Kansas City—Laboratory Results by Evidence Category and Crime Classification

Evidence Category	Crime Type	N*	Laboratory Results (%)					
			Identification	Negative Identification	Common Origin	Different Origin	Reconstructive	Inconclusive
Blood	Pers.	70	100	0	6	0	4	0
	Prop.	8	100	0	13	0	0	12
Semen	Pers.	44	75	23	0	0	2	7
	Prop.	0	—	—	—	—	—	—
Hair	Pers.	61	18	0	26	20	3	46
	Prop.	2	—	—	—	—	—	—
Firearms	Pers.	102	39	0	45	1	37	18
	Prop.	0	—	—	—	—	—	—
Toolmarks	Pers.	5	60	0	40	20	0	0
	Prop.	10	50	0	10	0	10	40
Prints	Pers.	115	2	0	46	14	0	46
	Prop.	72	0	0	7	10	0	83
Trace/transfer	Pers.	11	36	0	27	18	0	27
	Prop.	13	0	0	31	23	0	46
Drugs	Pers.	15	67	27	7	0	0	0
	Prop.	5	80	20	0	0	20	0
Flammables/explosives	Pers.	2	—	—	—	—	—	—
	Prop.	47	62	28	8	4	45	17
Impressions/patterns	Pers.	6	33	0	33	0	33	0
	Prop.	4	—	—	—	—	—	—

*Values where N < 5 cases are not computed.

Table XIII. Oakland—Laboratory Results by Evidence Category and Crime Classification

Evidence Category	Crime Type	N*	Laboratory Results (%)					
			Identification	Negative Identification	Common Origin	Different Origin	Reconstructive	Inconclusive
Blood	Pers.	60	65	8	40	8	0	13
	Prop.	3	—	—	—	—	—	—
Semen	Pers.	54	70	30	2	2	0	5
	Prop.	0	—	—	—	—	—	—
Hair	Pers.	12	25	8	67	8	0	0
	Prop.	5	—	—	—	—	—	—
Firearms	Pers.	120	41	1	48	12	14	14
	Prop.	5	50	0	20	0	20	20
Toolmarks	Pers.	0	—	—	—	—	—	—
	Prop.	1	—	—	—	—	—	—
Prints	Pers.	67	1	0	37	44	0	19
	Prop.	16	0	0	25	63	0	19
Trace/transfer	Pers.	1	—	—	—	—	—	—
	Prop.	15	13	0	33	7	0	47
Drugs	Pers.	9	56	44	0	0	0	0
	Prop.	2	—	—	—	—	—	—
Flammables/explosives	Pers.	0	—	—	—	—	—	—
	Prop.	0	—	—	—	—	—	—
Impressions/patterns	Pers.	9	33	0	11	22	33	11
	Prop.	6	0	0	50	50	0	0

*Values where N < 5 cases are not computed.

of positive identifications is close to the 70% mark in the other laboratories.

Although the number of hair submissions in Oakland is small ($N =$ 12), in two-thirds of the cases this evidence results in a conclusion of possible or probable common origin. The N of cases in Peoria and Kansas City with hair is about the same ($N = 60$). Common origin results develop in from one-quarter to one-third of the instances in which this evidence is submitted.

The percentage of submissions in which firearms evidence results in a statement of common origin is comparable in personal crimes from city to city; Peoria has the highest rate, 62%. Peoria also has the highest rate of toolmark cases in property crimes: 82% have a common origin result. None of the 21 toolmark cases sampled in Chicago result in a common origin finding. The Chicago toolmarks section examines many more cases than does Peoria. But, because it usually fails to receive a tool to compare with the toolmarks, examinations usually only yield information as to the type of tool that may have been used. This information may help the investigator subsequently to locate the proper suspect.

Peoria, once again, has the highest rate of trace/transfer evidence resulting in a common origin in both personal and property crimes. The Oakland samples include no trace evidence (glass, hair, fibers, etc.) in personal crimes and the Chicago sampling has too few to tabulate.

The presence of drug evidence in cases where other physical evidence is submitted is tabulated as well. Suspected drugs are identified as controlled substances between one-half and three-quarters of the time.

Impression and pattern evidence was found in a very small number of incidents in all cities; Peoria and Chicago have the most cases. This evidence has a high rate of positive outcome, with the results either demonstrating a common or different origin or, perhaps, helping to reconstruct the offense.

The final category included on the table is suspected accelerants and explosives. The rates of identification in Chicago and Kansas City (50–60%) are comparable. Suspected arson accelerants are very rarely examined in Peoria and Oakland and are subsequently not included in this review.

Questioned documents are not included in the tabulation because they are examined only by the Chicago Crime Laboratory. Chicago is the only one of the four facilities in the study having the capability of examining documents for the purpose of determining their authenticity and authorship (origin). A check of Chicago results reveals that in 16% of the incidents, a statement of common origin (conclusive, probable, or possible) is made. These are principally cases linking handwriting on a document (fraudulent check, credit card) to a specific individual. In another 24% of the cases, examiners are able to classify the make or model of a typewriter used to type a document or, possibly, to determine that some currency is counterfeit. In about half the cases sampled, however, no definitive results are reported.

The Value of Evidence: Resolving the Question of Association

Table XIV presents data that express the percentage of time in which the analysis of various categories of evidence resolve the question of possible association among suspects, victims, crime scenes, and instruments of the crime. Only those evidence categories that are commonly considered to have associative value are included in this table. Items such as drugs, flammables, explosives, and semen evidence are excluded because the standard laboratory procedure in these cases is primarily to identify the substance. Since the initiation of the study in 1980, most of the laboratories have begun programs to determine the blood group of the semen donor, which should enhance the associative power of this rape evidence.

An example of how the table may be read is as follows: blood evidence is examined in 93 cases in Peoria in which the purpose for submission is to associate persons, a person and a location, or possibly a person and an instrument of the crime. In 31% of these cases, bloodstain evidence either confirms or refutes this association.

Rather than comparing percentages for evidence categories between cities it is probably more useful to examine the relative rates of success

Table XIV. Percent of Time Laboratory Results Are Successful in Determining if Persons/Objects Are Associated with One Another

Evidence Category	Crime Type	Peoria N(%)	Chicago N(%)	Kansas City N(%)	Oakland N(%)
Blood	Personal	93(31)	76(33)	24(38)	53(36)
	Property	4(50)	26(8)	5(40)	3(0)
Hair	Personal	75(39)	6(50)	52(50)	11(36)
	Property	1(100)	0(0)	15(7)	0(0)
Fingerprints	Personal	48(65)	34(24)	151(48)	81(64)
	Property	18(61)	38(3)	156(7)	24(54)
Firearms/ toolmarks	Personal	104(86)	138(49)	112(70)	129(71)
	Property	33(70)	38(5)	9(22)	3(33)
Trace/ transfer	Personal	17(59)	2(100)	8(38)	1(100)
	Property	25(64)	3(33)	12(50)	15(53)

The percentages in this table are derived by dividing the number of times that laboratory results either associated *or* disassociated persons, weapons, tools, or scenes of crimes by the number of times evidence is submitted to the laboratory for that purpose (the N value).

enjoyed by various evidence categories across all jurisdictions. This approach reveals the following.

In personal crimes:

- firearms evidence is by far the category of evidence that yields the greatest success in resolving the question of association;

- bloodstain evidence is at the bottom of the rankings in three of the four cities in its ability to show a positive or negative association; and

- fingerprints rank high in comparison to most other evidence categories, placing either second or third in all cities.

In property crimes:

- trace evidence is successful in resolving the issue of association more than half the time;

- toolmarks associate tools with crime scenes from a high of 70% to a low of 5% of the time; and

- in contrast to their value in personal crimes, fingerprints have a much poorer record in associating and disassociating persons in property offenses.

Laboratory Results When Only Fingerprints Are Collected and Examined

In a very high percentage of burglary crime scene investigations, only fingerprints are gathered. Because these cases constitute one of the major activities of crime scene units and represent a significant fraction of all cases in which physical evidence is collected, they deserve special treatment. They have not been discussed up to this point because fingerprint identification is usually handled by a unit external to the crime laboratory. Information on cases involving fingerprints as the only category of physical evidence was collected in Peoria, Chicago, and Oakland. The sample was not collected in Kansas City because of recordkeeping limitations. Table XV compares the utilization of fingerprint evidence in three separate types of cases:

1. burglary/property crimes for which *only* fingerprints are collected (FP-Burg);
2. burglary/property crimes for which other physical evidence is examined in the crime laboratory (Ev-Burg); and

Table XV. Utilization of Fingerprint Evidence

City	Sample	Number of Cases	Average Number Collected	Percent of Cases with Fingerprint Collected	Average Number Analyzed	Percent of Cases with Fingerprint Analyzed
Peoria	FP-Burg	34	1.12	100	1.00	100
	Ev-Burg	62	2.03	32	1.56	26
	Ev-Other	219	2.79	32	1.84	21
Chicago	FP-Burg	42	1.00	100	1.00	100
	Ev-Burg	80	1.86	34	1.25	24
	Ev-Other	296	1.74	22	1.57	14
Oakland	FP-Burg	33	1.18	100	1.00	100
	Ev-Burg	43	2.07	53	1.20	40
	Ev-Other	229	4.77	49	1.45	29

3. other, non-burglary, crimes with physical evidence examined in the crime laboratory (Ev-Other).

The second and third categories of cases described above may or may not have had fingerprints collected in addition to the evidence examined in the laboratory.

In Table XV, the column giving the average number of physical evidence categories collected refers to the average number collected per case. The third row lists the percentage of cases in that group that have fingerprint evidence collected, so it follows that 100% of the FP-Burg group have fingerprint evidence collected. The fourth row, marked "analyzed," records the average number of physical evidence categories receiving scientific analysis per case. In the FP-Burg cases, only fingerprints have been examined, so the average is 1.00 in all cities. Finally, the last column gives the percentage of cases in each group that have fingerprint evidence examined.

This table clearly illustrates that crimes considered more serious than burglaries, specifically murder, rape, robbery, and assault, result in more evidence collection and laboratory analysis. Not only is more evidence collected in the more serious crimes (which has been shown previously in this chapter), but the quality of the evidence appears to be enhanced. As shown in Table XVI, when fingerprints are collected in the more serious crimes, standards are more likely to be collected as well. Also, the laboratory appears to be better able to reach a common origin result through the evidence analysis.

In Peoria, for example, the fingerprints of a suspect are compared with prints from a crime scene in just 32% of the burglary cases in which only fingerprints are collected. In burglaries, when other evidence is ex-

Table XVI. Results of Fingerprint Analyses

City	Sample	Number of Cases	Percent of Fingerprint with Both Evid & Stds	Percent of Fingerprint with Common Origin	Percent of Common Origin with Both Evid & Stds
	FP-Burg	34	32	24	75
Peoria	Ev-Burg	16	69	50	72
	Ev-Other	47	87	77	89
	FP-Burg	42	10	5	50
Chicago	Ev-Burg	19	16	16	100
	Ev-Other	40	25	23	90
	FP-Burg	33	42	3	7
Oakland	Ev-Burg	17	82	18	21
	Ev-Other	67	91	36	39

amined in the laboratory, fingerprint standards are available in 69% of the cases in which latent prints are recovered. In crimes other than burglary, fingerprint standards are available in 87% of the cases. One can see that as the rate of standards present increases, so does the rate of common origin fingerprint results (i.e., the latent print is matched with a particular person).

In Chicago, only 10% of the "fingerprint-only" burglaries have standards available. In other words, the prints of particular suspects are checked against the unknown latent fingerprints recovered in the field in only 10% of these crimes. This is the primary reason why fingerprints are matched with an individual only 5% of the time in these cases.

In Oakland, while latent prints are compared with suspect fingerprints in 42% of cases, they only match up 7% of the time (*see* the far right hand column). This result suggests that the quality of suspect names submitted to the fingerprint identification section in Oakland does not equal the quality of suspect names supplied in the other jurisdictions.

Summary

The principal conclusions that may be drawn from the data presented in this chapter are summarized as follows:

- Several characteristics of a criminal act are associated with the collection of evidence; among them: the type of offense, the interaction between suspect and the scene or victim, the seriousness of injury suffered by the victim, the location of the crime (residential versus nonresidential), the presence of witnesses, the status of identity of suspects, and the presence of supervisory police personnel at the crime scene.

- Biological fluids and firearms dominate as the primary evidence categories collected and analyzed in personal crimes, while fingerprints, trace evidence, and toolmarks are the leading categories of evidence examined in property crimes.

- The principal reason evidence is submitted to the laboratory, putting drug evidence aside, is to associate persons, weapons, tools, and locations with one another.

- On the average, many more categories of evidence are collected in personal crimes than in property crimes.

- Only a fraction of evidence collected from the field is analyzed; the highest percentage is examined in property crimes and the lowest in personal crimes.

- The jurisdiction that gathers the greatest quantity of evidence from the scenes of crimes (Oakland) also examines the fewest categories of evidence in those cases.

- The percentage of laboratory results leading to a statement of common origin is highest in personal crimes; on the other hand, property crimes return the highest number of different origin results.

- The Peoria jurisdiction has the greatest success in determining the origin of firearms, toolmarks, fingerprints, and trace evidence. Oakland has the greatest success in determining the origin of bloodstains and hair evidence. Chicago and Kansas City have the highest rates of identifying semen evidence in sexual crimes.

- Firearms, bloodstains, and toolmarks are the leading evidence categories in personal crimes that successfully resolve questions of association among persons and locations. Trace and toolmark evidence are the primary categories in property crimes that resolve the question of association.

- Fingerprint evidence is most successful in identifying persons when it is collected in conjunction with other evidence in nonburglary/property crime cases. It is successful the smallest percentage of the time when it is the only item of evidence gathered in property crimes.

Acknowledgments

This chapter is based on an article appearing in the final report "Forensic Evidence and the Police: The Effects of Scientific Evidence on Criminal Investigations."

This original research was supported under a grant from the National Institute of Justice and the Forensic Sciences Foundation. Points of view or opinions in this chapter are those of the author and do not necessarily represent the official position or policies of the Forensic Sciences Foundation or the National Institute of Justice.

The author acknowledges the contribution of Steven Mihajlovic, Research Associate of the Center for Research in Law and Justice, who provided invaluable assistance in the collection and analysis of data for this project.

APPENDIX A

Background Information on Various Attributes of Study Sites

Attribute	Peoria/Morton	Chicago	Kansas City	Oakland
Population	125,639	3,060,801	462,914	344,686
Crime index total	12,054	186,728	42,065	41,269
Index crime per capita (1000s)	95.9	61.0	90.9	119.7
Land area (square miles)	38	228	317	59
Laboratory established	1972	1930	1973	1944
Parent law enforcement agency	Peoria Police Department (crime scene unit) Illinois Department of Law Enforcement (crime laboratory)	Chicago Police Department	Kansas City Police Department	Oakland Police Department
Sworn personnel	218 (Peoria Police Department)	12,392	1,183	602
Index crimes per sworn officer	55:1	15:1	36:1	69:1
Criminal investigators	35 (Peoria P.D.)	1,268	204	126
Organizational placement of crime laboratory	Bureau of Scientific Services	Bureau of Technical Services	Bureau of Criminal Investigations	Bureau of Investigations

Background Information on Various Attributes of Study Sites—Continued

Attribute	Peoria/Morton	Chicago	Kansas City	Oakland
Scope of service	Regional	Municipal	Regional	Municipal
Crime laboratory caseload	2,697	25,600	10,926	5,364
Ratio of cases to examiner	300:1	512:1	840:1	766:1
Number of scientific staff	9(1)*	50	13(10)*	7(5)*
Ratio of sworn staff to (proportionate) scientific staff	24:1(218:1)	248:1	91:1(118:1)	86:1(120:1)
Parent police department budget	$4,315,530	$351,415,466	$35,826,402	$39,148,857
Crime laboratory budget (excludes crime scene search function)	—	$ 1,300,000 (approx.)	$ 275,290	$ 171,836
Ratio of laboratory budget to parent agency budget	—	0.4%	0.8%	0.4%
Crime scene function: organizational unit	Peoria P.D. Admin. Services	Crime Laboratory	Criminalistics Division	Patrol Division
Number of crime scene personnel	6	95	22	12
Index crimes per technician	2,009	1,966	1,912	3,439
Ratio sworn staff to technicians	36:1	130:1	54:1	50:1
Ratio of crime scene technicians to (proportionate) laboratory staff	0.67:1(6:1)	1.9:1	1.7:1(2.2:1)	1.7:1(2.4:1)

NOTE: Unless otherwise indicated, the information in this table describes the characteristics of the agencies and laboratories as they were in 1979.

*The number in parentheses refers to the proportionate number of scientific staff in the Morton and Kansas City regional laboratories examining cases from the Peoria and Kansas City jurisdictions. Approximately 10% of the Morton Regional Laboratory caseload is from Peoria and 80% of the Kansas City Regional Laboratory caseload is from Kansas City. This translates into 10% × 9 or 1 staff member in the Morton laboratory working Peoria cases and 80% × 13 or 10 staff members in the Kansas City laboratory working Kansas City Police Department cases. The Oakland Crime Laboratory staff of 7 includes 2 full-time fingerprint examiners; so to make the Oakland staffing level equivalent to the other laboratories, these 2 fingerprint examiners are excluded.

Background Information on Physical Evidence Examination Capabilities of Study Sites

Evidence	Peoria/Morton	Chicago	Kansas City	Oakland
Blood/alcohol	O	O	X	O
Comp. microscopy	X	X	X	X
Crime scene search	X	X	X	X
Drugs	X	X	X	X
Explosives	O	X	X	O
Fibers	X	X	X	X
Fingerprints	X	X	X	X
Flammables	X	X	X	X
Firearms	X	X	X	X
Glass	O	X	X	X
Gunshot residue	O	X	X	O
Hair	X	X	X	X
Paint	X	X	X	X
Polygraph	X	X	O	X
Questioned document	O	X	O	O
Serial number restoration	X	X	X	X
Serology	X	X	X	X
Soils/minerals	O	X	X	O
Toolmarks	X	X	O	X
Toxicology	X	O	X	O
Trace	X	X	X	X
Voiceprints	O	O	O	O

KEY: X, crime laboratory has examination capability; and 0, crime laboratory lacks examination capability.

Background Information on Reference Collections in Study Sites

Reference Collections (Standard/open case file)	Peoria/Morton	Chicago	Kansas City	Oakland
Laundry/dry cleaning	No/No	No/No	No/No	No/No
Tire	No/No	No/No	No/No	No/No
Auto paints	Yes/No	Yes/No	Yes/No	Yes/No
Hair	Yes/No	Yes/No	Yes/No	Yes/No
Fibers	Yes/No	Yes/No	Yes/No	Yes/No
Shoe prints	No/No	Yes/Yes	Yes/No	No/No
Instrumental	Yes/No	Yes/Yes	Yes/No	Yes/No
Threatening letters/bad checks	No/No	Yes/Yes	No/No	No/No
Bullets/cartridge cases	Yes/Yes	Yes/Yes	Yes/Yes	Yes/Yes
Fingerprints	No/No	Yes/Yes	No/No	Yes/Yes
Wood	No/No	No/No	Yes/No	Yes/No
Blood	Yes/No	Yes/No	Yes/Yes	Yes/No

INDEX

INDEX

Production Editor: Paula M. Bérard

Book and cover design and illustrations: Anne G. Bigler

Managing Editor: Janet S. Dodd

Typesetting: FotoTypesetters, Baltimore, MD, and Hot Type Ltd., Washington, DC

Printing and binding: Maple Press Company, York, PA